SpringerBriefs in Electrical and Computer Engineering

Series editors
Woon-Seng Gan, School of Electrical and Electronic Engineering, Nanyang Technological University, Singapore, Singapore
C.-C. Jay Kuo, University of Southern California, Los Angeles, CA, USA
Thomas Fang Zheng, Research Institute of Information Technology, Tsinghua University, Beijing, China
Mauro Barni, Department of Information Engineering and Mathematics, University of Siena, Siena, Italy

W0079193

SpringerBriefs present concise summaries of cutting-edge research and practical applications across a wide spectrum of fields. Featuring compact volumes of 50 to 125 pages, the series covers a range of content from professional to academic. Typical topics might include: timely report of state-of-the art analytical techniques, a bridge between new research results, as published in journal articles, and a contextual literature review, a snapshot of a hot or emerging topic, an in-depth case study or clinical example and a presentation of core concepts that students must understand in order to make independent contributions.

More information about this series at http://www.springer.com/series/10059

Xiaoming Chen

Massive Access for Cellular Internet of Things Theory and Technique

 Springer

Xiaoming Chen
Zhejiang University
Hangzhou, Zhejiang, China

ISSN 2191-8112   ISSN 2191-8120   (electronic)
SpringerBriefs in Electrical and Computer Engineering
ISBN 978-981-13-6596-6   ISBN 978-981-13-6597-3   (eBook)
https://doi.org/10.1007/978-981-13-6597-3

Library of Congress Control Number: 2019935144

This Springer imprint is published by the registered company Springer Nature Singapore Pte Ltd.
The registered company address is: 152 Beach Road, #21-01/04 Gateway East, Singapore 189721, Singapore

Preface

We are living in the era of the Internet of Things (IoT), which has affected and even changed our styles of work, study, and life in depth. Everyday, from the time we wake up, we are used to checking the weather forecast from the IoT. When we go to bed at night, the IoT is still measuring our sleeping information. Nowadays, we have been surrounded by various IoT devices at home, on the road, and in the office. All the time, a lot of new devices access the IoT network, and the IoT network becomes bigger and bigger. It is predicted that the number of IoT devices will reach to 20.4 billion by 2020.

Without doubt, to achieve the goal of IoT, the focus is on the Internet, but not the things. Especially, the IoT devices should be interconnected through a wireless mode. Currently, the IoT devices access various wireless networks mainly via the Zigbee, Bluetooth, and Wi-Fi techniques. However, these techniques only support moderate and small-range wireless access, e.g., in the building or campus. Yet, more and more IoT applications require a seamless access over a large range. For instance, we usually run with some wearable devices around the West Lake in Hangzhou. In this case, the traditional techniques cannot provide reliable wireless access for a massive number of IoT devices. It is clear that a promising solution is the use of the existing cellular networks, namely, cellular IoT. In this context, 3GPP (3rd Generation Partnership Project) made a specification for the cellular IoT in Release 13 in 2015. In particular, by making use of a variety of new radio techniques, e.g., massive MIMO, NOMA, mmWave, and new waveform, the 5G cellular network is expected to provide massive access for the cellular IoT with stringent QoS requirements, e.g., low latency, ultra-reliability, low power, and high mobility. Thus, the cellular IoT can satisfy the requirement of various IoT applications.

Even with new radio techniques, it is not a trivial task for the 5G to support massive access of the cellular IoT over limited radio spectrum. Especially, there are various different IoT scenarios, which may not have a unified solution. In this book, we aim to present some feasible solutions for a few typical cellular IoT scenarios from the viewpoints of both theory and technique. In Chap. 1, we introduce the characteristics of the cellular IoT and its key techniques for achieving efficient massive access. Next, Chap. 2 addresses the problem of massive access of the cellular IoT over stationary channels in the scenario of fixed devices. Then, Chap. 3 considers the cellular IoT over slowly time-varying fading channels in FDD (frequency division duplex) mode. The achievable rate is analyzed, and the corresponding performance optimization methods are provided. Then, the cellular IoT over slowly time-varying fading channel in TDD (time division duplex) mode is studied; a new fully non-orthogonal communication framework for massive access is proposed. Furthermore, the high-mobility scenario of the cellular IoT is discussed in Chap. 5, and a beamspace massive access technique is designed based on the available channel state information. Finally, we present a summary about massive access for the cellular IoT in 5G and beyond and point out the future research directions for further improving the overall performance of the cellular IoT in Chap. 6. It is sincerely expected that this book can provide useful insights for the design and optimization of the cellular IoT.

Hangzhou, China   Xiaoming Chen
August 2018

Contents

Chapter 1
Introduction

Abstract With the increasing development of IoT, a massive number of IoT devices are desired to access various wireless networks, so as to provide a variety of advanced applications in industry, agriculture, medicine, and environment. In order to satisfy differential performance requirements of the IoT applications with limited wireless resources, one key point is the design of efficient multiple access schemes. In this chapter, we first discuss the advantages of cellular IoT over the other IoT networks and the development trend of cellular IoT in 5G and beyond. Then, we give an overview of massive access techniques of the cellular IoT, which will be frequently applied and redesigned for different scenarios of the cellular IoT in the sequent chapters. Finally, we introduce the objective and content of this book.

1.1 Cellular IoT

We have come to the era of Internet of Things (IoT), which is so big and getting bigger [1]. As of 2017, there are 8.4 billion connected things in use all over the world. It is predicted that the number of IoT devices will reach to 20.4 billion by 2020 [2]. The increasing rate is quite amazing, and without of doubt it will be faster in the next decade. Thus, IoT will be internet of everything soon.

The increasing development of IoT is mainly due to the strong demands and wide applications [3–6]. Nowadays, as shown in Fig. 1.1, IoT has been applied in nearly all fields of society and life, e.g., industry, agriculture, traffic, and medicine. Every day, there are a large number of IoT devices around us, maybe at home, on the road, and in the office used for sensing, computing, control and communication. Actually, we hardly study and work without the IoT. In general, most of the IoT devices are some certain small-size and low-power nodes with simple function units. For different applications, the IoT devices are equipped with differential function units in sensing, computing, control and communication, so as to achieve the performance requirements in sensing precision, processing speed, power consumption, latency, security, reliability, and transmission rate [7].

X. Chen, *Massive Access for Cellular Internet of Things Theory and Technique*,
SpringerBriefs in Electrical and Computer Engineering,
https://doi.org/10.1007/978-981-13-6597-3_1

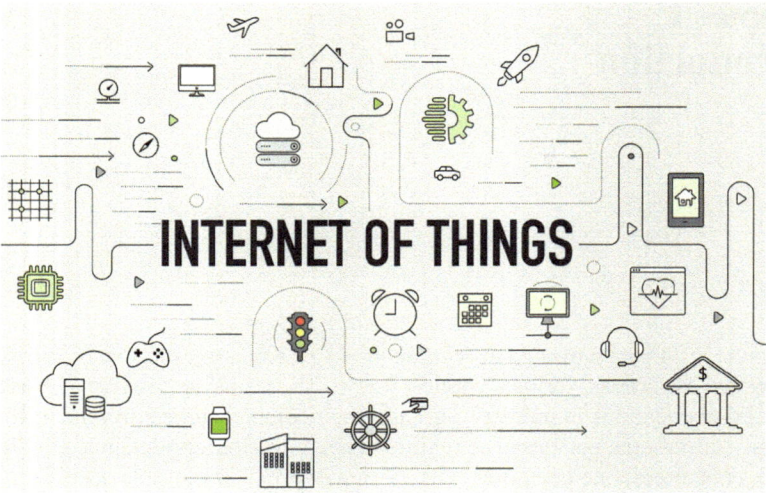

Fig. 1.1 The IoT applications

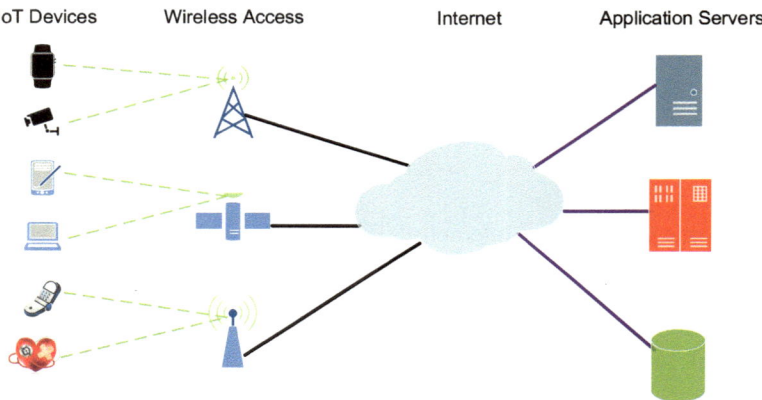

Fig. 1.2 A IoT network

The capability of a IoT device is quite limited, since it is usually a simple node for ease of deployment. In order to implement the functions of IoT, it is not enough to have a large number of simple devices. In other words, the key of IoT is not the thing, but is the internet. In general, a IoT network contains four components [7, 8], i.e., IoT devices, wireless access, internet, and application servers, cf. Fig. 1.2. Specifically, the IoT devices access the internet through various wireless access points, e.g., cellular BS, WiFi AP, and satellite, then exchange information with the application servers [9, 10]. Due to a massive number of IoT devices, the amount of exchanged information is enormous. As is well know that the nodes of core networks are connected through large-capacity optical fibers, hence the information can be exchanged in core networks at a high speed. Yet, the wireless access network is a main bottleneck limiting the performance of IoT, because the capacity of radio access interface is low. Especially, it is directly affected by the complicated wireless propagation environments, e.g., fading, noise and interference [11]. Therefore, the design of wireless access modes is a key of improving the overall performance of the IoT.

1.1.1 Why Is Cellular IoT

IoT has been applied in various fields, and thus there exist multiple wireless access techniques designed according to the characteristics and requirements of the IoT application scenarios [2]. In general, there are the following four main wireless access techniques:

1. Zigbee [12]: Zigbee is a wireless mesh network standard built on the IEEE 802.15.4 specifications. It works on the 2.4 GHz unlicensed radio spectrum, and the communication distance is about 10 m. The supportable maximum rate is 250 kbps. IoT devices can easily access the network via Zigbee gateways. Zigbee is an open, low-power, secure, global, and interoperable standard protocol, and is widely applied in the home automation or smart home.
2. Bluetooth [13]: Bluetooth is also an open and low-power wireless network standard, which operates on the 2.4 GHz frequency. The extended communication range of bluetooth is within 50 m, but the supportable rate is low. Thus, it can be adopted in smart home and industrial IoT.
3. WiFi [14]: WiFi, known as IEEE 802.11, is also a wireless technology standard on the 2.4 GHz band. It supports moderate-range and high-throughput wireless access. As a result, the WiFi devices have high energy consumption due to the severe interference on the unlicensed spectrum. WiFi has been widely used in home, campus, and building.
4. Cellular [15]: Cellular network is the biggest wireless network in the world, and has covered 90% of the world's population. It supports long-distance, high-throughput, low-latency, and ultra-reliable wireless communications.

In Table 1.1, we compare the four wireless access techniques from a variety of perspectives concerned by the IoT applications. It is seen that the cellular IoT has the advantages in all key performance metrics. Especially, the cellular network adopts a variety of advanced new radio (NR) techniques, e.g., massive multiple-input multiple-output (MIMO) and millimeter wave (mmWave) techniques, thus it is likely to support nearly all communication scenarios [16, 17]. Hence, the cellular IoT can provide a unified solution for various IoT applications.

Table 1.1 Comparison of four wireless access techniques		Zigbee	Blue	WiFi	Cellular
	Throughput	Moderate	Low	High	High
	Range	Short	Short	Moderate	Long
	Security	Moderate	Low	Moderate	High
	Power	Low	Low	High	Low
	Mobility	No	No	No	Yes
	Latency	Low	Low	Low	Low
	Cost	Low	Low	Low	Low

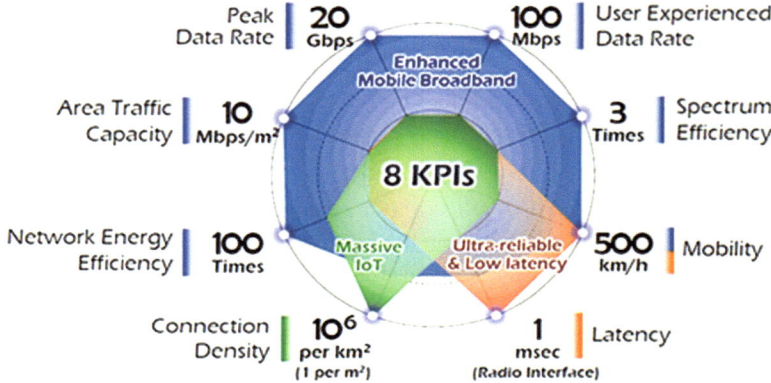

Fig. 1.3 The KPIs of 5G cellular networks

1.1.2   Cellular IoT in 5G and Beyond

The cellular network is evolving into the 5G, and the beyond 5G wireless network has been in research [18]. Compared to the previous cellular networks, e.g., 4G, the 5G network has more and stringenter key performance indicators (KPIs) [19], cf. Fig. 1.3. It is seen that the 5G cellular IoT is required to support very high connectivity density with a high energy efficiency. Thus, it can achieve massive IoT access with a low power consumption. Especially, 3GPP has made the specification of cellular IoT in the Release 13 in 2015, and hence cellular IoT will be a key component of the 5G network [20].

As shown in Fig. 1.4, the 5G cellular IoT should support a variety of differential IoT applications, from fixed monitor to high-mobility UAV, and from low-rate voice communication to real-time video transmission. To cover all IoT scenarios, the 5G cellular IoT adopts several novel techniques [21], i.e., long-term evolution for machines (LTE-M) and narrow band IoT (NB-IoT). LTE-M is suited for higher bandwidth or mobile and roaming applications, e.g., UAV supervision and vehicular equipments. Differently, NB-IoT suits low bandwidth, infrequent communications from relatively stationary devices, e.g., remote environment sensors for temperature, wind, pressure measurement.

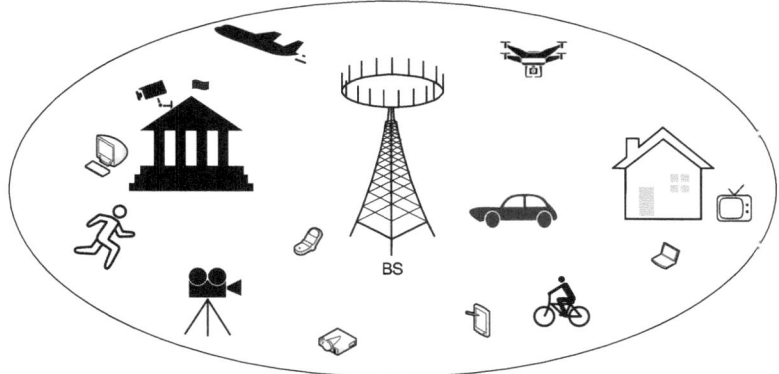

Fig. 1.4   A 5G cellular IoT network

1.2   An Overview of Massive Access for the Cellular IoT

The 5G cellular IoT has a great potential of supporting the access of a massive number of IoT with different QoS requirements. However, it is not a trivial task to achieve massive access over limited radio spectrum. Traditional multiple access techniques, e.g., frequency division multiple access (FDMA) and time division multiple access (TDMA), allocate an orthogonal time-frequency resource block to an user exclusively, resulting in a low spectral efficiency. Especially, with limited wireless resources, the above orthogonal multiple access (OMA) techniques cannot support massive access. In this context, non-orthogonal multiple access (NOMA) is widely recognized as an enabling technique of 5G cellular networks [22, 23]. In general, NOMA deploys superposition coding at the transmitter to realize spectrum sharing of a massive number of users and performs successive interference cancel (SIC) at the receiver to mitigate partial co-channel interference [24]. Yet, even with SIC, there still exists severe co-channel interference, especially in the scenario of massive access. As a result, the quality of the received signal might be unsatisfactory for the IoT applications. To solve this problem, it is desired to combine NOMA and interference cancellation techniques to improve the overall performance. Considering that multiple-antenna BS is a fundamental characteristic of the 5G cellular networks and it has a strong capability of interference mitigation, it is common to adopt multiple-antenna NOMA technique to realize massive access with QoS guarantees for the cellular IoT [25–27]. In what follows, we introduce the key steps of the multiple-antenna NOMA technique for the cellular IoT with massive connections.

1.2.1 CSI Acquisition

CSI availability at the BS is the precondition of designing massive access schemes for the cellular IoT. Generally speaking, there are three categories of CSI at the BS. The first one is instantaneous CSI. In the ideal case, the BS may obtain full instantaneous CSI if the channels vary very slowly [28]. For instance, the environment sensors are usually fixed, and then the channels remain unchanged during a long time. Thus, it is possible to obtain full CSI. However, if the IoT devices move, the channels may not vary so slowly. As a result, the BS has no time to obtain full CSI, but only partial CSI. In this case, there are usually two CSI acquisition methods. If the system operates in FDD mode, the CSI can be acquired through feedback from the devices to the BS [29]. Otherwise, in TDD mode, the BS obtains the CSI about the uplink channels through channel estimation, and takes it as the CSI about the downlink channels due to channel reciprocity [30]. Note that the two CSI acquisition methods consume a lot of wireless resources for CSI feedback or channel estimation. In the scenario of massive access, the resource consumption might be prohibitive. Thus, the conventional CSI acquisition methods should be redesigned according to the characteristics of massive access in the cellular IoT.

The second one is statistical CSI. For the high-mobility IoT devices, e.g., UAV and vehicular equipments, the associated channels vary very fast, resulting in a short channel coherent time. In other words, it is difficult for the BS to obtain the instantaneous CSI. Fortunately, since the statistical CSI, i.e., channel mean and variance, remains constant within a relatively long time. Moreover, statistical CSI can be easily obtained at the BS by averaging over channel realizations [31]. Hence, statistical CSI suits for the high-mobility IoT scenarios.

The third one is no CSI. In some special scenarios, the IoT devices cannot help the BS to obtain the CSI. Thus, the BS has no any information about the downlink channels. Thereby, the BS has to adopt a certain fixed transmission scheme independent of channel conditions [32].

1.2.2 User Clustering

User clustering is carried out to achieve a balance between system performance and computational complexity. As mentioned above, SIC is conducted at the devices to mitigate partial co-channel interference. Specifically, a device first decodes the interfering signals related to the devices with weaker channel gains, then subtracts the interfering signals in the received signal, finally demodulates the desired signal. Thus, the devices with the strongest channel gain needs to mitigate the interference from all the other devices, resulting in a very high computational complexity. Since the IoT devices are usually simple nodes, they may not afford the high computational complexity. To address this problem, the devices are partitioned into several clusters [33]. The SIC is only performed within a cluster, hence the complexity can be decreased significantly.

It is intuitive that user clustering has a great impact on the system performance. On the one hand, if all IoT devices are grouped into one cluster, it is possible to achieve the optimal performance, but the complexity of SIC is unbearable. On the other hand, if a cluster only contains one device, SIC is avoidable, but there might be not enough degrees of freedom to admit a massive number of IoT devices. In general, a cluster contains a few IoT device to balance the performance and the complexity. Since the multiple-antenna NOMA scheme is adopted for massive access, it makes sense to perform user clustering in the spatial domain. Specifically, the devices are partitioned into multiple clusters according to their spatial information, and the devices in a cluster share the same spatial beam, but the beams across the clusters are orthogonal of each other [34, 35]. For spatial user clustering, there are two basic principles. First, the devices in the same spatial direction are grouped into a cluster, which is beneficial to mitigate the inter-cluster interference by spatial beamforming. Second, the devices in a cluster should have distinct channel gains, so as to facilitate the SIC. In order to perform user clustering in the spatial domain, the BS should have CSI related to the devices. If the BS has full instantaneous CSI, it is likely to perform the optimal user clustering. However, since user clustering is a mixed integer programming problem, the optimal solution only can be obtained through exhaustive searching. If there is partial instantaneous CSI, the user clustering can be conducted according to the channel direction information and channel gains. In the scenario of statistical CSI at the BS, the devices with the same direction in the statistical sense can be arranged in a cluster. In the worst case that there is no CSI at the BS, the position information can be utilized to perform user clustering.

1.2.3 Superposition Coding

Superposition coding is used to realize efficient spectrum sharing for a massive number of IoT devices. In the user clustering-based multiple-antenna NOMA systems, superposition coding includes two steps. First, the signals in a cluster are weighted summed with the transmit powers [36]. Then, the superposition coded signals are further weighed summed by using the spatial beams [37]. In general, the superposition coding in a cluster, namely power allocation, is used to coordinate the intra-cluster interference, while the superposition coding across the clusters, namely spatial beamforming, is conducted for mitigating the inter-cluster interference. Thus, the design of superposition coding is equivalent to a joint optimization problem of power allocation and spatial beamforming.

The power allocation for massive access is not a trivial task, because the impacts of the powers related to the devices are coupling. Thus, it is difficult to obtain the optimal solution. To resolve the power allocation problem, it is necessary to decouple the transmit powers by using some convex approximation methods [38]. Moreover, even with a suboptimal performance, the fixed-ratio power allocation method can be applied for simplifying the implementation complexity while satisfying the requirement of channel gain distinction for facilitating the SIC [39].

The spatial beamforming is not a new topic for multiple-antenna systems. However, in the cellular IoT with massive connections, the implementation complexity is a critical issue. Thus, some simple beamforming schemes, e.g., match filtering (MF) and zero-forcing (ZF) beamforming, is commonly used. Especially, in the 5G network with a large-scale antenna array at the BS, MF and ZF can achieve the asymptotically optimal performance [40, 41]. It is worth pointing out that the performance of spatial beamforming depends on the CSI accuracy. Since the BS usually has partial CSI, some robust beamforming schemes are designed for guarantee the performance in the worst case.

1.2.4 Successive Interference Cancellation

SIC is performed at the IoT devices for further improving the quality of the received signal. Although SIC has been extensively studied in multiuser systems, there exists a new characteristic for the SIC in the cellular IoT. During the SIC, the devices should decode and cancel the interfering signals. As mentioned above, most of the IoT devices are simple nodes with a limited computation capability. As a result, the decoding error may occur, and the co-channel interference cannot be cancelled. The residual interference may further increase the error probability of the decoding of the sequent interfering signals, namely error propagation [42]. Finally, the imperfect SIC degrades the performance of the desired signal [43]. Thus, the design of massive access for the cellular IoT should take into account imperfect SIC.

In order to alleviate the impact of imperfect SIC in the cellular IoT, we first need to model the impact of imperfect SIC on the quality of the received signal, e.g., signal-to-interference-plus-noise ratio (SINR). However, since imperfect SIC is affected by multiple factors, such as transmit power, channel condition, and processing capability of the IoT devices, it seems difficult to build an accurate model for the imperfect SIC. In previous works, as a first attempt, a linear model is constructed to curve the impact of imperfect SIC [29, 30]. Given the model, it is possible to alleviate the effect through optimizing CSI acquisition, user clustering, and superposition coding.

1.3 Objective of This Book

The cellular IoT is developing with an unprecedented speed, and hence changes our work, study, and life. However, there still exist several challenging issues, e.g., battery capacity, network congestion, and information security, which significantly affect the further development of the cellular IoT [44, 45]. In general, these issues can be solved or partially solved by carefully designing the massive access techniques. For instance, a good massive access technique can effectively reduce the power consumption, relieve the network congestion, and enhance information security. Therefore, the design of massive access techniques is a core issue of the cellular IoT. In this context, this book aims to study the theories and techniques of massive access according to the characteristics of IoT applications, and thus provides useful insights for the design and optimization of the cellular IoT.

As discussed in the last section, all key steps of massive access in the cellular IoT are dependent of available CSI at the BS. In order to elaborate the theories and techniques more clearly, we provide a detailed massive access framework under each type of CSI. The main contents of this book are as follows:

In Chap. 2, we consider a cellular IoT where the IoT devices are fixed. Thus, the channels vary very slowly, and the BS is able to obtain full CSI about the downlink channels of the IoT devices. Based on full CSI, the IoT devices are partitioned into multiple clusters according to their channel direction and gain. On the one hand, the nearly same channel direction within a cluster is beneficial to mitigate the inter-cluster interference and enhance the quality of the received signal. On the other hand, the distinct channel gain in a cluster facilitates to perform SIC. Then, based on available CSI and user clustering, the BS carries out superposition coding on the signals to be transmitted. The superposition coding includes two steps, one is power allocation within the cluster, and the other is spatial beamforming across the clusters. The power allocation within the cluster can coordinate the intra-cluster interference, while spatial beamforming may mitigate the inter-cluster interference. Finally, the BS broadcasts the coded signal over the downlink channels, and the IoT devices perform SIC within a cluster. Due to energy and size constraints, the IoT devices have limited computational ability. Thus, the decoding error during the SIC is inevitable, resulting in residual interference after SIC. To alleviate the impact of imperfect SIC on the performance of massive access, we propose to jointly optimize the spatial beam and transmit power from the perspectives of maximizing the weighted sum rate and minimizing the total power consumption, respectively. Moreover, for further reducing the computational complexity, we provide two simplified massive access algorithms adopting ZF beamforming at the BS fixedly. It has been shown that the proposed massive access schemes can effectively alleviate the impact of imperfect SIC, and thus enhance the robustness of the cellular IoT.

In Chap. 3, we focus on a cellular IoT where the IoT devices move at a low speed. In such a scenario, the channels are slowly time-varying. In other words, the channels remain constant within a time slot, and fade over time slots. We consider the cellular IoT operates in FDD mode, such that CSI should be conveyed from the IoT devices to the BS through quantized feedback. To be specific, each IoT device first obtains the CSI about the downlink channel through channel estimation, and quantizes the CSI through selecting an optimal codeword from a predetermined codebook. Then, the IoT device conveys the index of the optimal codewords with a finite number of bits to the BS, and thus the BS can recover the quantized CSI from the same codebook. Since the BS only has partial CSI, it carries out user clustering according to the position information of the IoT devices. In specific, the IoT devices in the nearly same physical direction but different access distances are grouped into a cluster. Furthermore, the BS directly conducts superposition coding based on imperfect CSI. Similarly, the signals within a cluster are weighted summed with the transmit power as the weighted coefficient, and then the signals of all clusters are weighted summed with the ZF beam as the weighted coefficient. Once the IoT devices receive the coded signal, SIC is adopted within a cluster to cancel partial intra-cluster interference. The number of quantization bits for CSI conveyance from the IoT device to the BS determines the CSI accuracy and thus the performance of the cellular IoT, but it is impossible to allocate a large number of feedback bits to each devices in the scenario of massive access due to a finite capacity of the feedback link. In this context, we propose to allocate the feedback bits according to channel conditions and system parameters, so as to obtain the satisfactory CSI for all devices with limited feedback resource. Moreover, we advocate to optimize the transmit power and transmission mode for further improving the overall performance of the cellular IoT. Simulation results validate the effectiveness of the proposed massive access based on quantized codebooks.

In Chap. 4, we also consider a cellular IoT with low-mobility IoT devices. Differently, the cellular IoT operates in TDD mode, and thus the BS can obtain the CSI by channel estimation directly. Generally speaking, the IoT devices send training sequences over the uplink channels, and the BS acquires the CSI about the uplink channels via channel estimation. Due to channel reciprocity in TDD mode, the uplink CSI can be used as the downlink CSI. In order to obtain accurate CSI, the training sequences are usually orthogonal of each other. In order to guarantee pairwise orthogonal, the length of training sequences should be larger than the number of IoT devices. However, in the scenario of massive access, the length of training sequences may become a large proportion of a time slot, resulting in a short duration for information transmission. In a worst case, the training sequence is longer than the channel coherent time. As a result, the estimated CSI is outdated. To solve this challenge, we propose a fully non-orthogonal communication framework for massive access. First, the channel estimation is non-orthogonal. The IoT devices in a cluster share the same training sequence, and the training sequences across the clusters are orthogonal. The sharing of training sequence sharply decreases the required number of training sequence, and then the length of training sequences is reduced. However, the non-orthogonal channel estimation may lead to the decreasing of CSI accuracy, and the CSI accuracy within a cluster are coupled. Second, the user access is also non-orthogonal. For the proposed fully non-orthogonal communication framework, there exists severe co-channel interference during the both stages of channel estimation and user access, resulting in a severe performance degradation. In order to achieve a spectral-efficient massive access for the cellular IoT, we propose to optimize the transmit power of the devices at the stage of channel estimation and the transmit power of the BS at the stage of user access accordingly. Numerical simulations show that the proposed fully non-orthogonal communication framework can support massive access with finite wireless resources.

In Chap. 5, we concentrate on a high-mobility cellular IoT, where the IoT devices move fast. Thus, the channels also vary so fast that it is impossible to feed back the CSI or send training sequences in each time slot. Meanwhile, it is necessary to design low-complexity superposition coding algorithm for reducing the processing delay over fast time-varying fading channels. In this context, we propose a non-orthogonal beamspace massive access scheme. In other words, we utilize beamspace CSI, namely statistical CSI, to design the massive access scheme. Since statistical CSI remains constant during a relatively long time, it is particularly applicable in the system in the fast time-varying environment. According to the characteristics of the beamspace, the base beams are asymptotically orthogonal. Thus, the whole beamspace are partitioned into multiple orthogonal subspaces, and each subspace relates to a base beam. For simplifying the user clustering, we group the IoT devices in a subspace into a cluster. Since the IoT devices in a cluster are randomly distributed in the subspace, the base beam cannot effectively enhance the performance of all the devices in a cluster. To this end, we propose to construct non-orthogonal transmit beam for the devices. First, we give a full-space multiple-beam design algorithm, which constructs an independent transmit beam for each device with all base beams. To reduce the design complexity, we then present a partial space design algorithm, which constructs an independent transmit beam for each device with partial base beams. For further decreasing the required number of transmit beams, we provide a partial-space single-beam design algorithm, which constructs a transmit beam for the devices in the same cluster with partial base beams. Since the transmit beam is a linear combination of the base beams, the superposition coding method has a low computational complexity. It has been shown that the proposed non-orthogonal beamspace massive access schemes perform better than the baseline schemes.

Finally, in Chap. 6, we give a summary about massive access for the cellular IoT in 5G and beyond. Especially, we provide several future research directions for further improving the overall performance of the cellular IoT.

References

1. A. Al-Fuqaha, M. Guizani, M. Mohammadi, M. Aledhari, M. Ayyash, Internet of things: a survey on enabling technologies, protocols, and applications. IEEE Commun. Survs. Tuts **17**(4), 2347–2376 (2015)
2. M.R. Palattella, M. Dohler, A. Grieco, G. Rizzo, J. Torsner, T. Engel, L. Ladid, Internet of things in the 5G era: enablers, architecture, and business models. IEEE J. Sel. Areas Commun. **34**(3), 510–527 (2016)
3. L.D. Xu, W. He, S. Li, Internet of things in industries: a survey. IEEE Trans. Ind. Inform. **10**(4), 2233–2243 (2014)
4. A. Zanella, N. Bui, A. Castellani, L. Vangelista, M. Zorzi, Internet of things for smart cities. IEEE Int. Things J. **1**(1), 22–32 (2014)
5. H. Zhang, J. Li, B. Wen, Y. Xun, J. Liu, Connecting intelligent things in smart hospitals using NB-IoT. IEEE Internet Things J. **5**(3), 1550–1560 (2018)

6. Y. Li, X. Cheng, Y. Cao, D. Wang, L. Yang, Smart choice for the smart grid: narrowband internet of things (NB-IoT). IEEE Internet Things J. **5**(3), 1505–1515 (2018)
7. J. Lin, W. Yu, N. Zhang, X. Yang, H. Zhang, W. Zhao, A survey on Internet of things: architecture, enabling technologies, security and privacy, and application. IEEE Internet Things J. **4**(5), 1125–1142 (2017)
8. S. Cirani, L. Davoli, G. Ferrari, R. Léone, P. Medagliani, M. Picone, L. Veltri, A scalable and self-configuring architecture for service discovery in the internet of things. IEEE Internet Things J. **1**(5), 508–521 (2014)
9. M.R. Palattella, N. Accettura, X. Vilajosana, T. Watteyne, L.A. Grieco, G. Boggia, M. Dohler, Standardized protocol stack for the internet of (important) things. IEEE Commun. Survs. Tuts. **15**(3), 1389–1406 (2013)
10. A. Rajandekar, B. Sikdar, A survey of MAC layer issues and protocols for machine-to-machine communications. IEEE Internet Things J. **2**(2), 175–186 (2015)
11. X. Chen, H.-H. Chen, W. Meng, Cooperative communications for cognitive radio networks-from theory to applications. IEEE Commun. Survs. Tuts. **16**(3), 1180–1193 (2014)
12. L. Davoli, L. Belli, A. Cilfone, G. Ferrari, From micro to macro IoT: challenges and solutions in the integration of IEEE 802.15.4/802.11 and sub-GHz technologies. IEEE Internet Things J. **5**(2), 784–793 (2018)
13. A. Harris III, V. Khanna, G. Tuncay, R. Want, R. Kravets, Bluetooth low energy in dense IoT environments. IEEE Commun. Mag. **54**(12), 30–36 (2016)
14. P.K. Sharma, Y.-S. Jeong, J.H. Park, EH-HL: effective communication model by integrated EH-WSN and Hybrid LiFi/WiFi for IoT. IEEE Internet Things J. **5**(3), 1719–1726 (2018)
15. A.I. Sulyman, S.M.A. Oteafy, H.S. Hassanein, Expanding the cellular-IoT umbrella: an architectural approach. IEEE Wirel. Commun. **24**(3), 66–71 (2017)
16. V.W.S. Wong, R. Schober, D.W.K. Ng, L.-C. Wang, *Key Technologies for 5G Wireless Systems* (Cambridge University Press, Cambridge, 2017)
17. X. Chen, Z. Zhang, H.-H. Chen, On distributed antenna system with limited feedback precoding-opportunities and challenges. IEEE Wirel. Commun. **17**(2), 80–88 (2010)
18. J.G. Andrews, S. Buzzi, W. Choi, S.V. Hanly, A. Lozano, A.C.K. Soong, J.C. Zhang, What will 5G be? IEEE J. Sel. Areas Commun. **32**(6), 1065–1082 (2014)
19. C.-X. Wang, F. Haider, X. Gao, X.-H. You, Y. Yang, D. Yuan, H.M. Aggoune, H. Haas, S. Fletcher, E. Hepsaydir, Cellular architecture and key technologies for 5G wireless communication networks. IEEE Commun. Mag. **52**(2), 122–130 (2014)
20. 3GPP TR 45.820, Technical specification group GSM/EDGE radio access network; cellular system support for ultra-low complexity and low throughput internet of things (CIoT), Nov 2015
21. M. Agiwal, A. Roy, N. Saxena, Next generation 5G wireless networks: a comprehensive survey. IEEE Commun. Survs. Tuts **18**(3), 1617–1655 (2016)
22. L. Dai, B. Wang, Y. Yuan, S. Han, I. Chih-lin, Z. Wang, Non-orthogonal multiple access for 5G: solutions, challenges, opportunities, and future research trends. IEEE Commun. Mag. **53**(9), 74–81 (2015)
23. C. Zhong, X. Hu, X. Chen, D.W.K. Ng, Z. Zhang, Spatial modulation assisted multi-antenna non-orthogonal multiple access. IEEE Wirel. Commun. **25**(2), 61–67 (2018)
24. Z. Ding, Z. Yang, P. Fan, H.V. Poor, On the performance of non-orthogonal multiple access in 5G systems with randomly deployed users. IEEE Signal Process. Lett. **21**(12), 1501–1505 (2014)
25. M. Shirvanimoghaddam, M. Dohler, S.J. Johnson, Massive non-orthogonal multiple access for cellular IoT: potentials and limitations. IEEE Commun. Mag. **55**(9), 55–61 (2017)
26. M. Shirvanimogaddam, M. Condoluci, M. Dohler, S.J. Johnson, On the fundamental limits of random non-orthogonal multiple access in cellular massive IoT. IEEE J. Sel. Areas Commun. **35**(10), 2238–2252 (2017)
27. C. Zhong, X. Hu, X. Chen, D.W.K. Ng, Z. Zhang, Spatial modulation assisted multi-antenna non-orthogonal multiple access. IEEE Wirel. Commun. **25**(2), 61–67 (2018)

28. X. Chen, R. Jia, D.W.K. Ng, On the design of massive non-orthogonal multiple access with imperfect successive interference cancellation. IEEE Trans. Commun. (99), 1–1 (2018)
29. X. Chen, Z. Zhang, C. Zhong, D.W.K. Ng, Exploiting multiple-antenna for non-orthogonal multiple access. IEEE J. Sel. Areas Commun. **35**(10), 2207–2220 (2017)
30. X. Chen, Z. Zhang, C. Zhong, R. Jia, D.W.K. Ng, Fully non-orthogonal communication for massive access. IEEE Trans. Commun. **66**(4), 1717–1731 (2018)
31. J. Choi, On the power allocation for MIMO-NOMA systems with layered transmission. IEEE Trans. Wirel. Commun. **15**(5), 3226–3237 (2016)
32. Z. Ding, F. Adachi, H.V. Poor, The application of MIMO to non-orthogonal multiple access. IEEE Trans. Wirel. Commun. **15**(1), 537–552 (2016)
33. L. Xu, R. Collier, G.M.P. O'Hare, A survey of clustering techniques in WSNs and consideration of the challenges of applying such to 5G IoT scenarios. IEEE Internet Things J. **4**(5), 1229–1249 (2017)
34. Y. Liu, M. Elkashlan, Z. Ding, G.K. Karagiannidis, Fairness of user clustering in MIMO non-orthogonal multiple access systems. IEEE Commun. Lett. **20**(7), 1464–1468 (2016)
35. X. Chen, Z. Zhang, C. Zhong, R. Jia, On the design of massive access, in *Proceedings of IEEE WCSP*, Nanjing, China (2017), pp. 1–6
36. Z. Ding, R. Schober, H.V. Poor, A general MIMO framework for NOMA downlink and uplink transmission based on signal alignment. IEEE Trans. Wirel. Commun. **15**(6), 4438–4454 (2016)
37. X. Chen, R. Jia, D.W.K. Ng, The application of relay to massive non-orthogonal multiple access. IEEE Trans. Commun. **66**(11), 5168–5180 (2018)
38. C.-L. Wang, J.-Y. Chen, Y.-J. Chen, Power allocation for downlink non-orthogonal multiple access system. IEEE Wirel. Commun. Lett. **5**(5), 532–535 (2016)
39. Z. Yang, Z. Ding, P. Fan, N. Al-Dhahir, A general power allocation scheme to guarantee quality of service in downlink and uplink NOMA systems. IEEE Trans. Wirel. Commun. **15**(11), 7244–7257 (2016)
40. Z. Ding, H.V. Poor, Design of massive-MIMO-NOMA with limited feedback. IEEE Signal Process. Lett. **23**(5), 629–633 (2016)
41. J. Ma, C. Liang, C. Xu, P. Li, On orthogonal and superimposed pilot schemes in massive MIMO NOMA systems. IEEE J. Sel. Areas Commun. **35**(12), 2696–2707 (2017)
42. P. Li, R.C. de Lamare, R. Fa, Multiple feedback successive interference cancellation detection for multiuser MIMO systems. IEEE Trans. Wirel. Commun. **10**(8), 2434–2439 (2011)
43. S.M.R. Islam, N. Avazov, O.A. Dobre, K.-S. Kwak, Power-domain non-orthogonal multiple access (NOMA) in 5G: potentials and challenges. IEEE Commun. Survs. Tuts **19**(2), 721–742 (2017)
44. J. Xu, J. Yao, L. Wang, Z. Ming, K. Wu, L. Chen, Narrowband internet of things: evolutions, technologies, and open issues. IEEE Internet Things J. **5**(3), 1449–1462 (2018)
45. X. Chen, Z. Zhang, C. Zhong, D.W.K. Ng, R. Jia, Exploiting inter-user interference for secure massive non-orthogonal multiple access. IEEE J. Sel. Areas Commun. **36**(4), 788–801 (2018)

Chapter 2
Massive Access with Full Channel State Information

Abstract In this chapter, we consider a scenario where the wireless devices in the cellular IoT are fixed, such that CSI keeps unchanged during a relatively long time and the BS is capable of obtaining full CSI. In such a scenario, the benefits of spatial DoF offered by the multiple-antenna BS are exploited to support massive access. In particular, transmit beams and powers are jointly optimized to mitigate the intra-cluster and inter-cluster interference, so as to improve the overall performance. Specifically, we design the joint optimization algorithms from the perspectives of maximizing the weighted sum rate and minimizing the total power consumption, respectively. Moreover, in order to reduce the computational complexity, we design the massive access algorithms with ZF beamforming fixedly. Simulation results show that the proposed algorithms can achieve obvious performance gain over the baseline algorithms. Especially, the proposed algorithms are able to alleviate the impact of imperfect SIC on the performance of massive access.

2.1 Introduction

It has been nowadays a common viewpoint that the cellular IoT should adopt non-orthogonal multiple access (NOMA) techniques to realize massive access over limited radio spectrum [1, 2]. In particular, the NOMA techniques make use of superposition coding at the transmitter and successive interference cancellation (SIC) to mitigate the inter-user interference caused by non-orthogonal transmission, and thus improve the spectral efficiency [3]. However, in the context of massive access, a large number of user equipments (UEs) are simultaneously connected to a base station (BS), resulting in strong residual interference even after SIC. Moreover, the complexity of SIC might be prohibitive if there exist a large number of connections, especially in case that the UEs have limited computing capability. Thus, it is necessary to partition the UEs into several clusters, and SIC is only carried out within each cluster [4]. Unfortunately, user clustering may lead to extra inter-cluster interference. In other words, user clustering should be utilized together with some effective interference cancellation techniques. In [5], multicarrier NOMA

© The Author(s), under exclusive license to Springer Nature Singapore Pte Ltd. 2019
X. Chen, *Massive Access for Cellular Internet of Things Theory and Technique*,
SpringerBriefs in Electrical and Computer Engineering,
https://doi.org/10.1007/978-981-13-6597-3_2

schemes were proposed to avoid the inter-cluster interference by assigning an orthogonal carrier to an user cluster in the scenarios of perfect and imperfect channel state information (CSI), respectively. However, the multicarrier NOMA scheme may lead to a loss of spectral efficiency due to the orthogonal transmission across clusters. As is well known that multiple-antenna beamforming is a powerful and commonly used interference cancellation technique through spatial beamforming [6–8]. Considering the BSs in future wireless communication systems are almost equipped with multiple antennas, it is natural to combine NOMA and multiple-antenna beamforming to enhance the performance [9, 10]. By making use of spatial degrees of freedom of a multiple-antenna system, it is likely to separate user clusters in spatial domain [11]. Thus, the inter-cluster interference can be decreased or even cancelled without extra radio resources. It was shown in [12] that if the BS had full channel state information (CSI), zero-forcing beamforming (ZFBF) was able to completely mitigate inter-cluster interference in the case that the UEs were equipped with a single antenna each. In the case of multiple-antenna UEs, it is possible to mitigate the inter-cluster interference by jointly designing precoding and detection vectors, namely signal alignment [13]. Furthermore, the signal alignment scheme for multiple-antenna NOMA was extended to the multiple-cell scenario in [14].

Other than inter-cluster interference, user clustering based NOMA also suffer from intra-cluster interference. Thus, the design of multiple-antenna NOMA should make a balance between inter-cluster and intra-cluster interference cancellation [15]. It was reported in [16] that user clustering, beamforming, and power allocation were three effective methods to combat inter-cluster and intra-cluster interference for multiple-antenna NOMA downlink. However, it is a challenging issue to jointly design user clustering, beamforming, and power allocation schemes, since all optimization parameters are coupled. In [17], transmit beam and power were jointly optimization from the perspective of maximizing the weighted sum rate of a multiple-antenna NOMA downlink. For ease of design, the work [17] only considered a simple scenario, where each cluster contained two UEs. Similarly, a two-user cluster was considered in [18]. To further optimize the sum rate, a heuristic user clustering algorithm was provided before beamforming design and power allocation in [18]. It is clear that the two-user cluster is inefficient for exploiting the potential of NOMA. In order to support massive access, it is necessary to investigate the issue of joint design in a general multiple-antenna NOMA downlink.

A common assumption in previous related works is that the UE is capable of performing perfect SIC, and thus completely cancels the inter-user interference from the weaker signals [19, 20]. However, in practical systems, SIC is not a trivial task. Especially in the context of massive access, various wireless devices have distinct detection capabilities. For some simple devices, the decoding error might be inevitable during the procedure of SIC [21]. As a result, the interference cannot be cancelled completely, namely imperfect SIC, which decreases the quality of the received signal. Especially, the decoding error in preceding interference cancellation may affect the current interference cancellation, namely error propagation. Thus, imperfect SIC has a great impact on the performance of NOMA, especially in massive access systems. Imperfect SIC in NOMA systems has attracted the attention, but is still an open issue. As pointed out in [22], there is no prominent research that provides a mathematical understanding of the effect of imperfect SIC on NOMA schemes. In this chapter, we aim to alleviate the impact of imperfect SIC in a massive NOMA downlink system by jointly optimizing transmit beams and powers at the multiple-antenna BS. The contributions of this chapter are listed as follows:

1. We study the practical issue of imperfect SIC in massive NOMA systems. In particular, we model the imperfect SIC as a coefficient based on long-term measurement, which makes it convenient to quantitatively analyze and effective alleviate the impact of imperfect CSI.
2. We optimize transmit beam and power from the perspective of maximizing the weighted sum rate, and propose a dual iterative jointly design algorithm. Furthermore, we present a low-complexity design algorithm based on ZFBF at the BS.
3. We optimize the massive NOMA system in the sense of minimizing the total power consumption, and respectively present a joint beamforming and power design algorithm and a low-complexity power allocation algorithm for the BS.
4. We show that the proposed algorithms are applicable to a general massive NOMA downlink system through analyzing the impact of imperfect SIC, and point out that they are also able to address the problem of joint optimization in the case of perfect SIC.

The rest of this chapter is organized as follows: Sect. 2.2 gives a brief introduction of a cellular IoT, and presents a general massive NOMA framework. Section 2.3 focuses on the design of transmit beam and power joint optimization algorithms from the perspectives of both maximizing the weighted sum rate and minimizing the total power consumption. Section 2.4 provides some simulation results to validate the effectiveness of the proposed algorithms. Finally, Sect. 2.5 concludes the chapter.

Notations We use bold upper (lower) letters to denote matrices (column vectors), $(\cdot)^H$ to denote conjugate transpose, $E[\cdot]$ to denote expectation, $\| \cdot \|$ to denote the L_2-norm of a vector, and $| \cdot |$ to denote the absolute value.

2.2 System Model and Problem Formulation

Let us consider a cellular IoT, where the IoT devices are nearly fixed. A base station (BS) with N_t antennas broadcasts messages to K single antenna UEs, cf. Fig. 2.1. Note that the number of UEs in the cellular IoT might be very large. In order to achieve a balance between system performance and computational complexity, the UEs are grouped into several clusters in spatial domain. In specific, the UEs in the same direction but with distinctive propagation distances are grouped into a cluster, and the UEs in a cluster share a transmit beam. On the one hand, the same direction

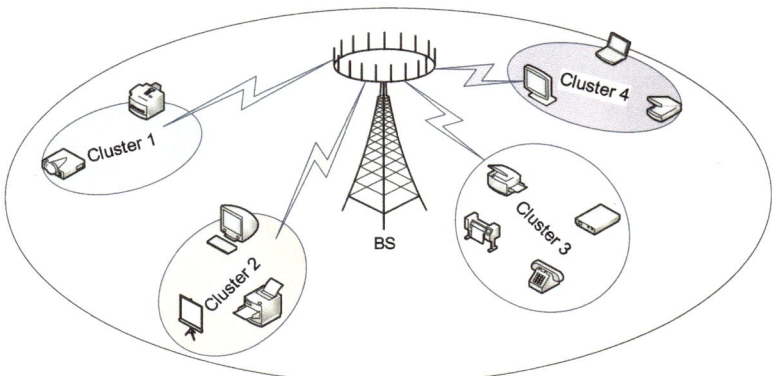

Fig. 2.1 A massive model of the cellular IoT

makes the beam aligned to all the UEs in such a cluster, which is beneficial to enhance the channel gain. On the other hand, a large gap of propagation distance enables SIC at the UEs more exact. Without loss of generality, we assume that the K UEs are partitioned into M clusters, and the mth cluster contains N_m UEs. For ease of description, we ues $UE_{m,n}$ to denote the nth UE in the mth cluster, and use $\mathbf{h}_{m,n}$ to denote the channel from the BS to the $UE_{m,n}$. It is assumed that the channels remain constant during one time slot, but independently fade over time slots. In general, CSI can be directly obtained by estimating the uplink channels in time division duplex (TDD) systems due to channel reciprocity. While in frequency division duplex (FDD) systems, CSI should be estimated at the UEs, and then is conveyed to the BS. In this chapter, we consider a scenario where the IoT devices are fixed, such that the BS is capable of obtaining full CSI by estimation or feedback at the beginning of each time slot.

In order to support massive access over limited radio spectrum, we adopt NOMA as the multiple access scheme. At first, the BS performs superposition coding on the signals to be transmitted. Different from traditional NOMA, the proposed NOMA for massive access needs two rounds of superposition coding. Generally speaking, in the first round, the BS constructs the transmit signal x_m for the mth cluster as follow:

$$x_m = \sum_{n=1}^{N_m} \sqrt{\alpha_{m,n}} s_{m,n}, \qquad (2.1)$$

where $s_{m,n}$ is the Gaussian distributed signal of unit norm for the nth UE in the mth cluster, and $\alpha_{m,n}$ is the intra-cluster power allocation factor with the following constraint:

$$\sum_{n=1}^{N_m} \alpha_{m,n} \leq 1. \qquad (2.2)$$

Note that intra-cluster power allocation is used to coordinate the intra-cluster interference, and thus improve the performance of downlink NOMA systems. Then, in the second round, the BS constructs the total transmit signal $\mathbf{x}$ as below:

$$\mathbf{x} = \sum_{m=1}^{M} \mathbf{w}_m x_m. \tag{2.3}$$

Finally, the BS broadcasts the signal $\mathbf{x}$ over the downlink channels. Without loss of generality, we consider the received signal $y_{m,n}$ at the nth UE in the mth cluster, which is given by

$$y_{m,n} = \mathbf{h}_{m,n}^H \mathbf{x} + n_{m,n}$$

$$= \underbrace{\mathbf{h}_{m,n}^H \mathbf{w}_m \sqrt{\alpha_{m,n}} s_{m,n}}_{\text{Desired signal}} + \underbrace{\mathbf{h}_{m,n}^H \mathbf{w}_m \sum_{i=1,i\neq n}^{N_m} \sqrt{\alpha_{m,i}} s_{m,i}}_{\text{Intra-cluster interference}}$$

$$+ \underbrace{\mathbf{h}_{m,n}^H \sum_{j=1,j\neq m}^{M} \mathbf{w}_j \sum_{i=1}^{N_j} \sqrt{\alpha_{j,i}} s_{j,i}}_{\text{Inter-cluster interference}} + \underbrace{n_{m,n}}_{\text{AWGN}}, \tag{2.4}$$

where $n_{m,n}$ is additive white Gaussian noise (AWGN) with variance σ^2. Note that the second term at the right side of Eq. (2.4) is the intra-cluster interference caused by non-orthogonal transmission, which can be reduced by performing SIC at the UEs. It is assumed that the effective channel gains in the mth cluster have a descending order, namely

$$|\mathbf{h}_{m,1}^H \mathbf{w}_m|^2 \geq \cdots \geq |\mathbf{h}_{m,N_m}^H \mathbf{w}_m|^2. \tag{2.5}$$

Thereby, according to the principle of NOMA, in the mth cluster, the jth UE can always successively decode the ith UE's signal $\forall j < i$, if the jth UE can decode its own signal. As such, the nth UE can subtract the interference from the $(n + 1)$th to N_mth UEs in its received signal $y_{m,n}$ before decoding its desired signal $s_{m,n}$. If SIC is perfectly carried out, the UE can completely cancel the intra-cluster interference from the UEs with weaker effective channel gains. However, in practical systems, due to the hardware limitation of mobile terminal, SIC might be imperfect. In other words, there exists the residual interference from the weaker users [23, 24]. As a result, the post-SIC signal and signal-to-interference-plus-noise ratio (SINR) at the UE$_{m,n}$ are given by

$$y'_{m,n} = \mathbf{h}_{m,n}^H \mathbf{w}_m \sqrt{\alpha_{m,n}} s_{m,n} + \mathbf{h}_{m,n}^H \mathbf{w}_m \sum_{i=1}^{n-1} \sqrt{\alpha_{m,i}} s_{m,i} + \sqrt{\eta_{m,n}} \mathbf{h}_{m,n}^H \mathbf{w}_m \sum_{i=n+1}^{N_m} \sqrt{\alpha_{m,i}} s_{m,i}$$

$$+\mathbf{h}_{m,n}^H \sum_{j=1,j\neq m}^{M} \mathbf{w}_j \sum_{i=1}^{N_j} \sqrt{\alpha_{j,i}} s_{j,i} + n_{m,n}, \tag{2.6}$$

and

$$\gamma_{m,n} = \frac{|\mathbf{h}_{m,n}^H \mathbf{w}_m|^2 \alpha_{m,n}}{|\mathbf{h}_{m,n}^H \mathbf{w}_m|^2 \sum_{i=1}^{n-1} \alpha_{m,i} + \eta_{m,n}|\mathbf{h}_{m,n}^H \mathbf{w}_m|^2 \sum_{i=n+1}^{N_m} \alpha_{m,i} + \sum_{j=1,j\neq m}^{M} |\mathbf{h}_{m,n}^H \mathbf{w}_j|^2 + \sigma^2}. \tag{2.7}$$

respectively, where $\eta_{m,n}$ denotes the coefficient of imperfect SIC at the $UE_{m,n}$, which can be obtained by long-term measurement.[1] Interestingly, it is found that for the nth UE in the mth cluster, the intra-cluster power allocation in the other clusters dose not affect its SINR.

As seen in (2.7), it is known that transmit beam $\mathbf{w}_m$ and power allocation $\alpha_{m,n}$ jointly determine the performance of massive access systems in the presence of imperfect SIC. In the following, we focus on the joint design of $\mathbf{w}_m$ and $\alpha_{m,n}$ for alleviating the impact of imperfect SIC.

2.3 Design of Massive Access System

In this section, we aim to jointly design transmit beamforming and powers for a general massive access system in the presence of imperfect SIC. As is well known, the maximization of weighted sum rate and the minimization of the total power consumption are two commonly used design objectives for multiuser systems. Therefore, we also design the massive access system from the two perspectives, respectively.

---

[1] By measuring a large number of samples in a long training time, the residual interference can be accurately approximated using a Gaussian distribution due to the central limit theorem, and the variance is a function of the received power [21]. Then, by comparing the powers of the residual interference and the received signal, the coefficient $\eta_{m,n}$ can be obtained.

2.3.1 *Weighted Sum Rate Maximization Design*

The design with the goal of maximizing the weighted sum rate of a multiple-antenna downlink NOMA system based on user clustering can be described as the following optimization problem:

$$J_1 : \max_{\mathbf{w}, \alpha} \sum_{m=1}^{M} \sum_{n=1}^{N_m} \theta_{m,n} R_{m,n}$$

$$\text{s.t. C1} : \sum_{m=1}^{M} \|\mathbf{w}_m\|^2 \leq P_{\max},$$

$$\text{C2} : \sum_{n=1}^{N_m} \alpha_{m,n} \leq 1, \forall m,$$

$$\text{C3} : \alpha_{m,n} \leq \alpha_{m,n+1}, n \in [1, N_m - 1],$$

where $R_{m,n} = \log_2(1 + \gamma_{m,n})$ is the achievable rate (in b/s) of the nth UE in the mth cluster, $\theta_{m,n} > 0$ denotes the priority, and $P_{\max}$ is the maximum transmit power budget at the BS. $\mathbf{w} = \{\mathbf{w}_1, \cdots, \mathbf{w}_M\}$ and $\alpha = \{\alpha_{1,1}, \cdots, \alpha_{M,N_M}\}$ are the collection of transmit beams and power allocation factors, respectively. Note that C1 represents the power allocation between the clusters, C2 denotes the power allocation among a cluster, and C3 is used to facilitate SIC in a cluster.

Since the optimization variables $\mathbf{w}$ and α are coupled in the objective function, J_1 is in general nonconvex. Thus, it is difficult to directly obtain the optimal solutions. To solve this problem, we transform the objective function according to the following lemma:

Lemma 1 *The received SINR $\gamma_{m,n}$ and the minimum mean squared error (MSE) $e_{m,n}$ between the transmit and receive signals have the following equivalent relation:* $1 + \gamma_{m,n} = e_{m,n}^{-1}$.

Proof Please refer to Appendix A.   □

Therefore, the objective function of J_1 is reduced as

$$\min_{\mathbf{w}, \alpha} \sum_{m=1}^{M} \sum_{n=1}^{N_m} \theta_{m,n} \log_2(e_{m,n}). \tag{2.8}$$

However, the objective function (2.8) is still not convex. Note that (2.8) aims to minimize a function of minimum MSE, which is equivalent to minimizing a function of MSE for a given MMSE receiver. In other words, the optimization objective in (2.8) can be transformed as

$$\min_{\mathbf{w},\mathbf{v},\boldsymbol{\alpha}} \sum_{m=1}^{M} \sum_{n=1}^{N_m} \theta_{m,n} \log_2(\mathrm{MSE}_{m,n}), \qquad (2.9)$$

where $\mathbf{v} = \{v_{1,1}, \cdots, v_{M,N_M}\}$ is the collection of the receivers, and $\mathrm{MSE}_{m,n}$ is the MSE related to the nth UE in the mth cluster. The sum of logarithmic function hinders us to further solve this problem. Similar to [25] and [26], we can replace the logarithmic function with the following term:

$$\min_{\mathbf{w},\mathbf{v},\boldsymbol{\alpha},\boldsymbol{\beta}} \sum_{m=1}^{M} \sum_{n=1}^{N_m} \theta_{m,n} (\beta_{m,n} \mathrm{MSE}_{m,n} - \log_2(\beta_{m,n})), \qquad (2.10)$$

where $\boldsymbol{\beta} = \{\beta_{1,1}, \cdots, \beta_{M,N_M}\}$ is a collection of auxiliary variables. Note that the objective function (2.10) achieves the minimum value only when $\beta_{m,n} = \mathrm{MSE}_{m,n}^{-1}, \forall m, n$. In such a case, the optimization objectives (2.9) and (2.10) are equivalent. Hence, the optimization problem J_1 is changed as

$$J_2: \quad \min_{\mathbf{w},\mathbf{v},\boldsymbol{\alpha},\boldsymbol{\beta}} \sum_{m=1}^{M} \sum_{n=1}^{N_m} \theta_{m,n} (\beta_{m,n} \mathrm{MSE}_{m,n} - \log_2(\beta_{m,n}))$$

$$\text{s.t. C1, C2, C3,}$$

According to the definition of $\mathrm{MSE}_{m,n}$ in (2.27), it is known that J_2 is not a joint convex function of $\mathbf{w}, \mathbf{v}, \boldsymbol{\alpha}, \boldsymbol{\beta}$, but it is a convex function of each optimization variable. Thus, we can apply the following property $\inf_{x,y} f(x, y) = \inf_x \tilde{f}(x)$ with $\tilde{f}(x) = \inf_y f(x, y)$ to solve J_2 [28]. Specifically, we optimize one variable by fixing the others. The four variables are iteratively optimized until they approach a stationary point. First, for the auxiliary variable $\beta_{m,n}$, it is always equal to $\mathrm{MSE}_{m,n}^{-1}$ in the sense of maximizing the weighted sum rate. Second, for the MMSE receiver, we have a closed-form solution $v_{m,n} = \sqrt{\alpha_{m,n}} \mathbf{w}_m^H \mathbf{h}_{m,n} \mathbf{U}_{m,n}^{-1}$. Since transmit beam $\mathbf{w}_m$ and power allocation factor $\alpha_{m,n}$ have multiple linear constraint conditions, they can be optimized by the Lagrange multiplier method. In specific, we construct the Lagrange function of J_2 as Algorithm 1:

$$\mathscr{L}_1(\mathbf{w}, \boldsymbol{\alpha}) = \sum_{p=1}^{M} \sum_{q=1}^{N_p} \theta_{p,q} (\beta_{p,q} \mathrm{MSE}_{p,q} - \log_2(\beta_{p,q})) + \mu \left(\sum_{p=1}^{M} \|\mathbf{w}_p\|^2 - P_{\max} \right)$$

$$+ \sum_{p=1}^{M} \omega_p \left(\sum_{q=1}^{N_p} \alpha_{p,q} - 1 \right) + \sum_{p=1}^{M} \sum_{q=1}^{N_m-1} \varsigma_{p,q} (\alpha_{p,q} - \alpha_{p,q+1}), \quad (2.11)$$

Algorithm 1 Massive access design for weighted sum rate maximization

---

Step 1: Initialize the parameters by letting $\mathbf{w}_m = \sqrt{\frac{P_{\max}}{M}}[1, 0, \cdots, 0]^H$, $\alpha_{m,n} = \frac{1}{N_m}$, $\forall m, n$;

Step 2: Set $v_{m,n} = \sqrt{\alpha_{m,n}}\mathbf{w}_m^H \mathbf{h}_{m,n} \mathbf{U}_{m,n}^{-1}$, $\beta_{m,n} = \mathrm{MSE}_{m,n}^{-1}$, $\mu = 1$, $\omega_m = 1$, $\varsigma_{m,n} = 1$, and $\varsigma_{m,n+1} = 1$;

Step 3: Update $\mathbf{w}_m$ according to (2.12), update $\alpha_{m,n}$ according to (2.13), and update μ, ω_m, $\varsigma_{m,n}$, and $\varsigma_{m,n+1}$ by the gradient method. Redo Step 3 until μ, ω_m, $\varsigma_{m,n}$ and $\varsigma_{m,n+1}$ are converged;

Step 4: Go to Step 2 until the sum rate is converged;

Step 5: Output $\mathbf{w}_m$ and $\alpha_{m,n}$.

---

where $\mu \geq 0$, $\omega_p \geq 0$, and $\varsigma_{p,q} \geq 0$, $\forall p, q$ are the Lagrange multipliers of C1, C2, and C3, respectively. Then, by making use of KKT conditions, we have

$$\mathbf{w}_m = \left(\sum_{p=1}^{M}\sum_{q=1}^{N_p}\sum_{n=1}^{N_m}\theta_{p,q}\beta_{p,q}v_{p,q}^H v_{p,q}\kappa_{m,n}^{p,q}\alpha_{m,n}\mathbf{h}_{p,q}\mathbf{h}_{p,q}^H + \mu\mathbf{I}\right)^{-1}\sum_{n=1}^{N_m}\sqrt{\alpha_{m,n}}\theta_{m,n}\beta_{m,n}\mathbf{h}_{m,n}v_{m,n}^H, \tag{2.12}$$

and

$$\alpha_{m,n} = \left(\frac{\theta_{m,n}\beta_{m,n}\mathrm{Re}(v_{m,n}\mathbf{h}_{m,n}^H\mathbf{w}_m)}{\sum_{p=1}^{M}\sum_{q=1}^{N_p}\theta_{p,q}\beta_{p,q}v_{p,q}^H v_{p,q}\kappa_{m,n}^{p,q}\left|\mathbf{h}_{p,q}^H\mathbf{w}_m\right|^2 + c_{m,n}}\right)^2, \tag{2.13}$$

respectively, where $c_{m,n} = \omega_m + \varsigma_{m,n} - \varsigma_{m,n+1}$, and $\mathrm{Re}(x)$ denotes the real component of complex number x. It is found that $\alpha_{m,n}$ is designed in the scale of a UE, while $\mathbf{w}_m$ is designed in the scale of a cluster. Thus, $\mathbf{w}_m$ is a weighted sum based on the parameters of the UEs in the cluster. In summary, the weighted sum rate maximization NOMA design is described as Algorithm 1.

Remarks The parameter $\kappa_{m,n}^{p,q}$ represents the impact of imperfect SIC on the design of massive NOMA systems. In Algorithm 1, if $\kappa_{m,n}^{p,q} = 0$, $\forall m = p$ and $n > q$, it becomes the case of perfect SIC. Thus, Algorithm 1 is a general design method for a multiple-antenna NOMA downlink system for all possible $\kappa_{m,n}^{p,q}$.

Note that Algorithm 1 requires two layers of iterations. The inner layer iteratively chooses μ, ω_m, $\varsigma_{m,n}$ and $\varsigma_{m,n+1}$ for computing $\mathbf{w}_m$ and $\alpha_{m,n}$ for given $v_{m,n}$ and $\beta_{m,n}$, while the outer layer iteratively updates $v_{m,n}$ and $\beta_{m,n}$. If we consider the iteration has converged when the change is less than δ, then the total number of iterations can be approximated as $\left(\log_2 \delta\right)^2$ [27]. In order to reduce the computational complexity, we propose a simplified massive NOMA design method based on ZFBF. Specifically, we design the transmit beam $\mathbf{w}_m'$ with unit norm in the null space of the channels related to the UEs in the other clusters, and only optimize the transmit power $P_{m,n}$ for maximizing the weighted sum rate. First, we introduce the design of ZFBF for the massive NOMA systems. For the design of the transmit beam $\mathbf{v}_m$, we first construct a complementary matrix as below:

$$\mathbf{H}_m = [\mathbf{h}_{1,1}, \mathbf{h}_{1,2}, \cdots, \mathbf{h}_{m-1,N_{m-1}}, \mathbf{h}_{m+1,1}, \cdots, \mathbf{h}_{M,N_M}]. \tag{2.14}$$

Then, we obtain the null space of the complementary matrix $\mathbf{H}_m$ as follows:

$$\boldsymbol{\Pi}_m = \mathbf{I} - \mathbf{H}_m \left(\mathbf{H}_m^H \mathbf{H}_m \right)^{-1} \mathbf{H}_m^H. \tag{2.15}$$

In order to improve the performance of the UE with the weakest channel gain, we design $\mathbf{w}_m'$ by projecting $\mathbf{h}_{m,N_m}$ on $\boldsymbol{\Pi}_m$, namely

$$\mathbf{w}_m' = \frac{\boldsymbol{\Pi}_m \mathbf{h}_{m,N_m}}{\| \boldsymbol{\Pi}_m \mathbf{h}_{m,N_m} \|}. \tag{2.16}$$

It is worth pointing out that we can project any UE's channel or a weight sum of all UEs' channel on $\boldsymbol{\Pi}_m$ to design $\mathbf{w}_m'$. Thus, we have

$$\mathbf{h}_{p,q}^H \mathbf{w}_m' = 0, \forall p \neq m. \tag{2.17}$$

In the case of ZFBF, the received SNR $\gamma_{m,n}$ at the nth UE in the mth cluster is given by

$$\gamma_{m,n}' = \frac{|\mathbf{h}_{m,n}^H \mathbf{w}_m'|^2 P_{m,n}}{|\mathbf{h}_{m,n}^H \mathbf{w}_m'|^2 \sum_{i=1, i \neq n}^{N_m} \kappa_{m,i}^{m,n} P_{m,i} + \sigma^2}, \tag{2.18}$$

where $P_{m,n}$ is the transmit power for the signal of the nth UE in the mth cluster. Thus, the maximization of weighted sum rate based on ZFBF can be described as the following optimization problem:

$$J_3: \max_{\mathbf{P}} \sum_{m=1}^{M} \sum_{n=1}^{N_m} \theta_{m,n} R_{m,n}'$$

$$\text{s.t. C4}: \sum_{m=1}^{M} \sum_{n=1}^{N_m} P_{m,n} \leq P_{\max},$$

$$\text{C5}: \; P_{m,n} \leq P_{m,n+1}, \forall m, \forall n \in [1, N_m - 1],$$

where $R_{m,n}' = \log(1 + \gamma_{m,n}')$ and $\mathbf{P} = \{P_{1,1}, \cdots, P_{M,N_M}\}$. Similarly, C5 is used to facilitate SIC at the UEs. Although J_3 reduces one optimization variable compared to J_1, the reminding optimization variable $P_{m,n}$ are still coupled in the objective function. Thus, it is difficult to obtain the optimal solution directly. To solve this problem, we adopt the sequential convex approximation (SCA) method to transform the original objective function to a series of solvable functions [28]. Specifically, we leverage a lower bound of logarithmic function to approximate the rate function. Mathematically, the approximation can be expressed as

$$R'_{m,n} = \log(1 + \gamma'_{m,n}) \geq a_{m,n} \log(\gamma'_{m,n}) + b_{m,n}, \tag{2.19}$$

where $a_{m,n} = \frac{\tilde{\gamma}'_{m,n}}{1+\tilde{\gamma}'_{m,n}}$ and $b_{m,n} = \log(1 + \tilde{\gamma}'_{m,n}) - \frac{\tilde{\gamma}'_{m,n}}{1+\tilde{\gamma}'_{m,n}} \log(\tilde{\gamma}'_{m,n})$ with $\tilde{\gamma}'_{m,n} \geq 0$ being an auxiliary variable equal to $\gamma'_{m,n}$ in the last iteration. In this case, the objective function of J_3 can be approximated as

$$\max_{\mathbf{P}} \sum_{m=1}^{M} \sum_{n=1}^{N_m} \theta_{m,n} \left(a_{m,n} \log \left(|\mathbf{h}^H_{m,n} \mathbf{w}'_m|^2 P_{m,n} \right) \right.$$
$$\left. -a_{m,n} \log \left(|\mathbf{h}^H_{m,n} \mathbf{w}'_m|^2 \sum_{i=1,i\neq n}^{N_m} \kappa^{m,n}_{m,i} P_{m,i} + \sigma^2 \right) + b_{m,n} \right). \tag{2.20}$$

Let $P_{m,n} = \exp(Q_{m,n})$, then the objective function (2.20) can be further transformed as

$$\max_{\mathbf{P}} \sum_{m=1}^{M} \sum_{n=1}^{N_m} \theta_{m,n} \left(a_{m,n} \log \left(|\mathbf{h}^H_{m,n} \mathbf{w}'_m|^2 \right) + Q_{m,n} \right.$$
$$\left. -a_{m,n} \log \left(|\mathbf{h}^H_{m,n} \mathbf{w}'_m|^2 \sum_{i=1,i\neq n}^{N_m} \kappa^{m,n}_{m,i} \exp(Q_{m,i}) + \sigma^2 \right) + b_{m,n} \right). \tag{2.21}$$

Since the log-sum-exp function is convex [28], (2.21) is a convex function. Additionally, the transformed constraint C4 $\sum_{m=1}^{M} \sum_{n=1}^{N_m} \exp(Q_{m,n}) \leq P_{\max}$ is also convex. Therefore, for a given $\tilde{\gamma}'_{m,n}, \forall m, n$, the solution of $Q_{m,n}$ can be obtained by making use of KKT conditions as below:

$$\theta_{m,n} - \sum_{q=1,q\neq n}^{N_m} \frac{\theta_{m,q} a_{m,q} |\mathbf{h}^H_{m,q} \mathbf{w}'_m|^2 \kappa^{m,q}_{m,n} \exp(Q_{m,n})}{|\mathbf{h}^H_{m,q} \mathbf{w}'_m|^2 \sum_{i=1,i\neq q}^{N_m} \kappa^{m,q}_{m,i} \exp(Q_{m,i}) + \sigma^2}$$
$$- \nu \exp(Q_{m,n}) - \xi_{m,n} \exp(Q_{m,n}) + \xi_{m,n-1} \exp(Q_{m,n}) = 0, \tag{2.22}$$

where $\nu \geq 0$ and $\xi_{m,n} \geq 0, \forall m, \forall n \in [1, N_m - 1]$ are the Lagrange multipliers related to C4 and C5, and it can be iteratively updated by the gradient method. If n is out of bound in (2.22), then $\xi_{m,n} = 0$. Then, update $\gamma'_{m,n}$ according to the

---

Algorithm 2 Simplified massive access design for weighted sum-rate maximization

Step 1: Initialize the parameters by letting $P_{m,n} = \frac{P_{\max}}{K}$, $\forall m, n$;

Step 2: Design ZFBF $\mathbf{w}'_m$ according to (2.16);

Step 3: Update $\tilde{\gamma}'_{m,n} = \dfrac{|\mathbf{h}^H_{m,n}\mathbf{w}'_m|^2 P_{m,n}}{|\mathbf{h}^H_{m,n}\mathbf{w}'_m|^2 \sum\limits_{i=1,i\neq n}^{N_m} \kappa^{m,n}_{m,i} P_{m,i}+\sigma^2}$, $\alpha_{m,n} = \dfrac{\tilde{\gamma}'_{m,n}}{1+\tilde{\gamma}'_{m,n}}$, $b_{m,n} = \log(1 + \tilde{\gamma}'_{m,n}) -$

$\dfrac{\tilde{\gamma}'_{m,n}}{1+\tilde{\gamma}'_{m,n}} \log(\tilde{\gamma}'_{m,n})$, $\nu = 1$, and $\xi_{m,n} = 1$;

Step 4: Obtain $Q_{m,n}$ by solving the equality (2.22);

Step 5: Update ν and $\xi_{m,n}$ by the gradient method, and go to Step 4 until ν and $\xi_{m,n}$ are converged;

Step 6: Go to Step 3 until $Q_{m,n}$ is converged;

Step 7: Output $\mathbf{w}'_m$ and $P_{m,n} = \log(Q_{m,n})$.

---

solution of $Q_{m,n}$ in the last iteration. The final solution of transmit power at the BS is obtained until the iteration converges. Hence, the simplified massive NOMA design can be summarized as Algorithm 2.

Note that according to the property of the SCA method, the weighted sum rate is increased after each iteration. Thus, Algorithm 2 is always converged due to the existence of the upper bound of sum rate.

Remarks Although Algorithm 2 can reduce the computational complexity, it needs more spatial degrees of freedom to design the beams compared to Algorithm 1. In specific, Algorithm 1 requires the BS to have M antennas, while Algorithm 2 needs at least $\max\limits_{1\leq m\leq M} \left(\sum\limits_{j=1,j\neq m}^{M} N_j \right) + 1$ antennas at the BS. This is because the transmit beam in Algorithm 1 is designed in the scale of cluster, but the transmit beam in Algorithm 2 must be in the null space of the channels of $\max\limits_{1\leq m\leq M} \left(\sum\limits_{j=1,j\neq m}^{M} N_j \right)$ UEs in the other clusters. In summary, for a given K, it is possible to relax the requirement on the number of antennas at the BS by reducing the number of cluster M for both Algorithms 1 and 2. However, reducing the number of clusters may lead to a high complexity of SIC. Thus, it is desired to choose a proper number of clusters for the design of massive NOMA systems.

2.3.2 Total Power Consumption Minimization Design

The incredible cost on power usage and the significant environment impact push wireless communications to meet various quality of service (QoS) requirements with the minimum power consumption. In this context, we intend to design the massive NOMA schemes from the perspective of minimizing the total power consumption. As mentioned above, the total power consumption at the BS is directly determined by transmit beams. For simplicity of design, we design a massive NOMA scheme with fixed-proportion power allocation within each cluster. Following previous related work on power allocation [9], we set $\alpha_{m,n} = \frac{n}{\sum_{i=1}^{N_m} i}$ fixedly

for balancing system performance and SIC implementation. Thus, the total power consumption minimization design with a constraint on the minimum rate can be described as the following optimization problem:

$$J_4 : \min_{\mathbf{w}} \sum_{m=1}^{M} \|\mathbf{w}_m\|^2$$

$$\text{s.t. C6} : \ R_{m,n} \geq r_{m,n}^0, \forall m, n,$$

where $r_{m,n}^0 > 0$ is a required minimum rate for meeting the QoS requirement. In general, J_4 is not convex due to the coupling of optimization variables in C6. To solve this problem, we reformulate J_4 as the following semi-definite programming (SDP) problem:

$$J_5 : \min_{\mathbf{W}} \sum_{m=1}^{M} \mathrm{tr}\left(\mathbf{W}_m\right) \mathrm{tr}\left(\mathbf{H}_{m,n}\mathbf{W}_m\right)$$

$$\text{s.t. C7} : \ \alpha_{m,n}\mathrm{tr}\left(\mathbf{H}_{m,n}\mathbf{W}_m\right) - \gamma_{m,n}^0 \sum_{i=1,i\neq n}^{N_m} \kappa_{m,i}^{m,n}\alpha_{m,i}$$

$$-\gamma_{m,n}^0 \sum_{j=1,j\neq m}^{M} \mathrm{tr}\left(\mathbf{H}_{m,n}\mathbf{W}_j\right) - \gamma_{m,n}^0\sigma^2 \geq 0,$$

where $\mathbf{H}_{m,n} = \mathbf{h}_{m,n}\mathbf{h}_{m,n}^H$, $\mathbf{W}_m = \mathbf{w}_m\mathbf{w}_m^H$, $\gamma_{m,n}^0 = \exp(r_{m,n}^0)$, and $\mathbf{W} = \{\mathbf{W}_1, \cdots, \mathbf{W}_M\}$. It is clear that J_5 is a convex function with respect to $\mathbf{W}$. Hence, one can derive the optimal solution $\mathbf{W}_m^*$ of J_5 by some optimization softwares, i.e, CVX [29]. If the rank of $\mathbf{W}_m^*$ is one, it is easy to obtain the solution $\mathbf{w}_m^*$ to J_4 by eigenvalue decomposition (EVD). Unfortunately, it is difficult to guarantee that $\mathbf{W}_m^*$ is always rank-one. Thus, $\mathbf{w}_m^*$ is in general obtained by some approximation methods, i.e., principle eigenvector [28]. In summary, the design of massive NOMA for minimizing the total power consumption can be described as Algorithm 3.

Algorithm 3 needs to use the approximation method to obtain the transmit beams, which may result in performance loss. In order to avoid this problem and further decrease the computational complexity, we design a massive NOMA scheme with ZFBF at the BS. With the same transmit beams $\mathbf{w}_m'$ in (2.16), the design of massive NOMA for minimizing the total power consumption is reduced to the following power allocation problem:

Algorithm 3 Massive NOMA design for total power consumption minimization

---

Step 1: Initialize the parameters by letting $\mathbf{w}_m = \sqrt{\frac{P_{\max}}{M}}[1, 0, \cdots, 0]^H$ and $\alpha_{m,n} = \frac{n}{\sum_{i=1}^{N_m} i}$, $\forall m, n$;

Step 2: Compute $\mathbf{W}_m$ by solving J_5 with CVX;

Step 3: Obtain $\mathbf{w}_m$ based on $\mathbf{W}_m$ by the approximation method;

Step 4: Output $\mathbf{w}_m$ and $\alpha_{m,n}$.

Algorithm 4 Simplified massive NOMA design for total power consumption minimization

---

Step 1: Initialize the parameters by letting $P_{m,n} = \frac{P_{\max}}{K}$, $\forall m, n$;

Step 2: Design ZFBF $\mathbf{w}'_m$ according to (2.16);

Step 3: Compute $P_{m,n}$ by solving the LP problem with CVX;

Step 4: Output $\mathbf{w}'_m$ and $P_{m,n}$.

$$J_6 : \min_{\mathbf{P}} \sum_{m=1}^{M} \sum_{n=1}^{N_m} P_{m,n}$$

$$\text{s.t. C5, C8} : R'_{m,n} \geq r^0_{m,n}, \forall m, n.$$

Without loss of generality, C8 can be reformulated as

$$|\mathbf{h}^H_{m,n}\mathbf{w}'_m|^2 P_{m,n} - \gamma^0_{m,n}|\mathbf{h}^H_{m,n}\mathbf{w}'_m|^2 \sum_{i=1,i\neq n}^{N_m} \kappa^{m,n}_{m,i} P_{m,i} - \gamma^0_{m,n}\sigma^2 \geq 0. \qquad (2.23)$$

Thus, J_6 is transformed as a linear programming (LP) problem, which can be effectively solved by some optimization softwares. Also, we provide the algorithm of the simplified design of massive NOMA for minimizing the total power consumption as Algorithm 4.

Moreover, for the simplified design with the purpose of minimizing the total power consumption, we have the following proposition:

Proposition 1 *The total power consumption is minimized only when all users achieve the minimum rate.*

Proof Please refer to Appendix B.

According to Proposition 1, the transmit power $P_{m,n}$ can be expressed as

$$P_{m,n} = \frac{\gamma^0_{m,n}\left(|\mathbf{h}^H_{m,n}\mathbf{w}'_m|^2 \sum_{i=1,i\neq n}^{N_m} \kappa^{m,n}_{m,i} P_{m,i} + \sigma^2\right)}{|\mathbf{h}^H_{m,n}\mathbf{w}'_m|^2}. \qquad (2.24)$$

Remarks It is also found that the parameter $\kappa_{m,i}^{m,n}$ determines the impact of imperfect SIC on the design of multiple-antenna NOMA downlink for minimizing the total power consumption. Note that Algorithm 4 is also applicable to the case of perfect SIC by letting $\kappa_{m,i}^{m,n} = 0, \forall i > n$. Especially, if SIC is perfect, it is possible to obtain the exact expression of transmit power. In specific, the transmit power for the first UE with the strongest effective channel gain in the mth cluster is give by

$$P_{m,1} = \frac{\gamma_{m,1}^0 \sigma^2}{|\mathbf{h}_{m,1}^H \mathbf{w}_m'|^2}. \tag{2.25}$$

Then, the transmit powers for the other UEs can be computed sequently according to the descending order of effective channel gains.

2.4 Numerical Results

In this section, we evaluate the performance of the proposed massive access schemes in several different scenarios. Without extra specification, we set $N_t = 64$, $K = 48$, $M = 12$, $N_m = N = 4$, $\eta_{m,n} = \eta = 0.05$, $\sigma^2 = 1$ and $\theta_{m,n} = 1, \forall m, n$. Moreover, we use SNR (in dB) to denote the ratio of transmit power at the BS and the noise variance.

Firstly, we show the performance gain of NONA over OMA in the context of massive access. In specific, we compare the sum rate of the proposed Algorithm 1, Algorithm 2 and a time division multiple access (TDMA) algorithm. As seen in Fig. 2.2, the NOMA scheme, especially Algorithm 1, performs much better than the OMA scheme (namely TMDA) in the whole SNR region. Thus, the proposed NOMA scheme has a great potential of supporting massive access over limited radio spectrum. For the proposed NOMA scheme, Algorithm 1 achieves a higher sum rate at low SNR, but Algorithm 2 performs better at high SNR. This is because the massive access system is noise limited when SNR is low, while it is interference limited as SNR increases. Algorithm 2 is capable of cancel the inter-cluster interference completely, and thus achieves a better performance at high SNR. In other words, if the SNR is high, one can utilize Algorithm 2 to realize simple and efficient massive access. However, it is worth pointing out that Algorithm 2 requires more antennas at the BS than Algorithm 1.

Secondly, we check the impact of the number of BS antennas N_t on the sum rate of Algorithm 1, cf. Fig. 2.3. For a given SNR, the sum rate of Algorithm 1 improves as the number of BS antennas increases. This is because the UEs can obtain more array gains for improving the performance. It is noticed that even with a small number of BS antennas, e.g., $N_t = 16$, Algorithm 1 can achieve a high sum rate. Thus, Algorithm 1 has a strong capability of supporting massive access with a limited number of BS antennas. Moreover, it is found that the sum rate will be asymptotically saturated as the SNR increases, since the massive NOMA system is

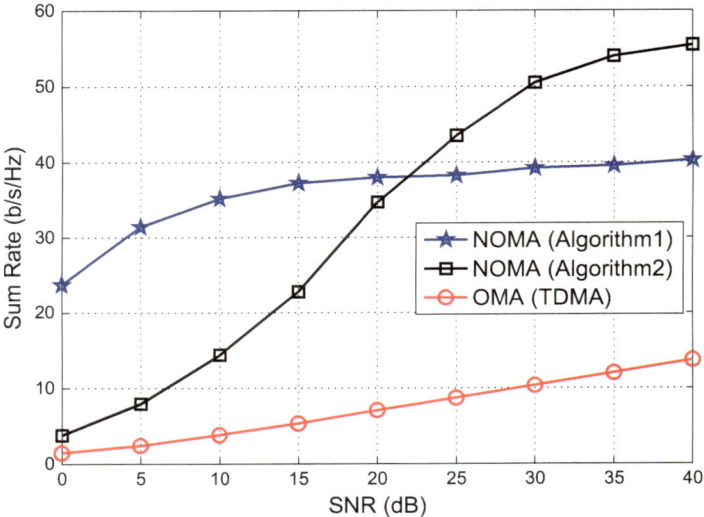

Fig. 2.2 Performance comparison of different massive access algorithms

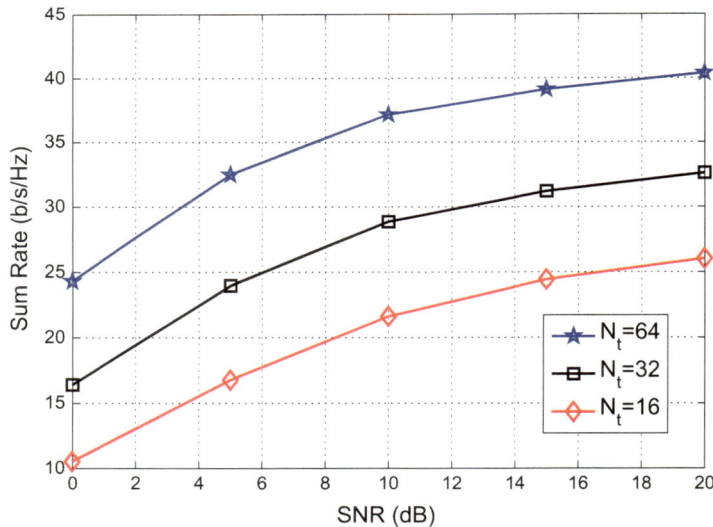

Fig. 2.3 Impact of the number of BS antennas N_t on the performance of Algorithm 1

interference limited at high SNR. In order to further improve the performance at high SNR, a feasible method is to increase the number of BS antennas. In fact, the BS in 5G systems will be equipped with a large-scale antenna array. Thus, it is easy to improve the performance of massive NOMA schemes by increasing the number of BS antennas.

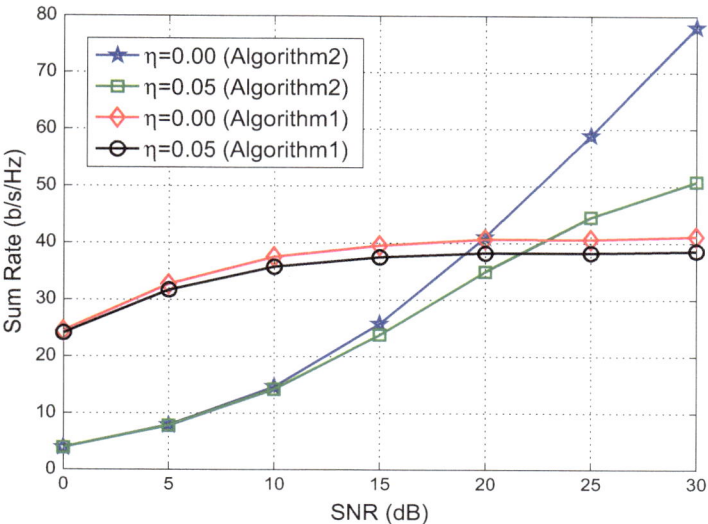

Fig. 2.4 Comparison of the capability of alleviating the impact of imperfect SIC

Figure 2.4 investigates the capability of the proposed massive NOMA schemes for alleviating the impact of imperfect SIC. For Algorithm 1, the performance loss due to imperfect SIC is slight in the whole SNR region, thus it is able to effectively eliminate the impact of imperfect SIC. However, for Algorithm 2, since it fixes the transmit beams, it has a weak capability of alleviating the impact of imperfect SIC. Thus, the performance loss increases as the SNR increases.

Then, we examine the power consumption of different massive access schemes for meeting the requirement on the minimum SINR. In Fig. 2.5, we compare four massive access schemes, namely TDMA, fixed NOMA with ZFBF $\mathbf{w}'_m$ in (2.16) and power allocation factor $\alpha_{m,n} = \frac{n}{\sum_{i=1}^{N_m} i}$, the proposed Algorithms 3 and 4. It is seen that Algorithm 3 consumes the smallest power in all scenarios. There is nearly a fixed performance gap between fixed NONA and Algorithm 4, since they utilize the same ZFBF. TDMA is very sensitive to the required minimum SINR. As the required minimum SINR increases, the total power consumption sharply increases, and it is much larger than that of the other schemes. Thus, the proposed massive NOMA schemes, especially Algorithm 3, can significantly reduce the power consumption.

Next, we show the effect of the number of UEs N in a cluster on the total power consumption. As seen in Fig. 2.6, for both Algorithms 3 and 4, the total power consumption increases as the number of UEs in a cluster increases. This is because there are more residual intra-cluster interference after SIC if the number of UEs in a cluster is large. Thus, more power should be consumed to meet the minimum SINR requirement. However, a small number of UEs in a cluster means that the BS should designed more transmit beams, resulting in a higher computational complexity.

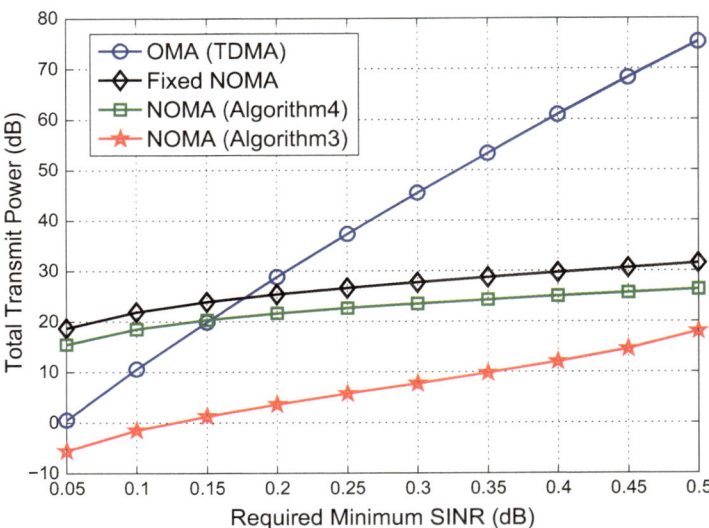

Fig. 2.5 Comparison of transmit power consumption of different massive access schemes for meeting the QoS requirements

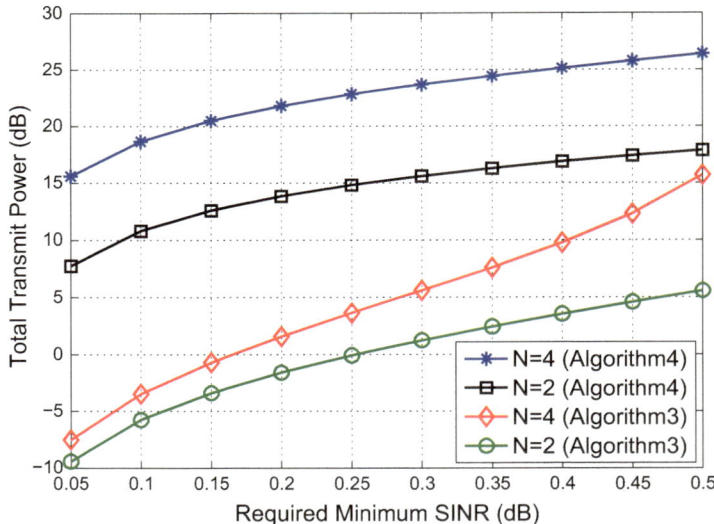

Fig. 2.6 Influence of the number of UEs in a cluster on the total power consumption

Moreover, if there are a small number of UEs in a cluster, it is difficult to support massive access with a finite number of BS antennas. Thus, it is desired to choose a proper number of UEs in a cluster.

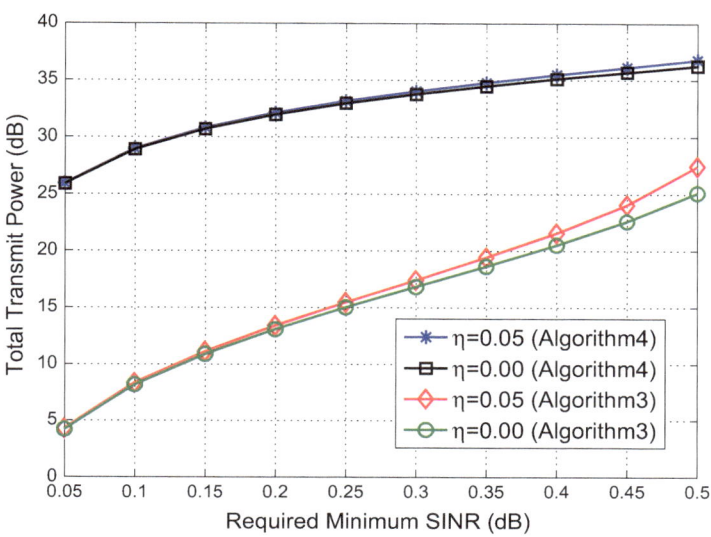

Fig. 2.7 Effect of imperfect SIC on the total power consumption

Finally, we check the impact of imperfect SIC on the power consumption of
the proposed Algorithms 3 and 4. It is intuitive that imperfect SIC leads to a high
residual intra-cluster interference, resulting in a large power consumption. However,
it is found in Fig. 2.7 that there is only a negligible improvement on the power
consumption for both Algorithms 3 and 4 when imperfect SIC occurs. Thus, the
both algorithms have strong capability of alleviating the impact of imperfect SIC.

2.5 Conclusion

In this chapter, we exploit the benefits of multiple-antenna NOMA to support
massive access over limited radio spectrum. For the critical issue of imperfect SIC
in practical NOMA systems, we propose to jointly optimize transmit beam and
power to alleviating the impact of imperfect SIC. Especially, we design the joint
optimization algorithms from the perspectives of maximizing the weighted sum rate
and minimizing the total power consumption, respectively. Extensive simulation
results validate the effectiveness of the proposed massive NOMA schemes, and also
reveal the effects of system parameters on the performance.

Appendix A   Proof of Lemma 1

According to the received signal $y_{m,n}$, the mean squared error (MSE) at the nth user in the mth cluster can be expressed as

$$\text{MSE}_{m,n} = \text{E}\left[(v_{m,n}y'_{m,n} - s_{m,n})(v_{m,n}y'_{m,n} - s_{m,n})^H\right], \tag{2.26}$$

where $v_{m,n}$ denotes the receiver at the nth user in the mth cluster. Substituting (2.6) into (2.26), $\text{MSE}_{m,n}$ can be expressed as

$$\text{MSE}_{m,n} = v_{m,n}\left(\sum_{j=1}^{M}\sum_{i=1}^{N_j}\kappa_{j,i}^{m,n}\alpha_{j,i}\mathbf{h}_{m,n}^H\mathbf{w}_j\mathbf{w}_j^H\mathbf{h}_{m,n} + \sigma^2\right)v_{m,n}^H - \sqrt{\alpha_{m,n}}\mathbf{w}_m^H\mathbf{h}_{m,n}v_{m,n}^H$$

$$-\sqrt{\alpha_{m,n}}v_{m,n}\mathbf{h}_{m,n}^H\mathbf{w}_m + 1 \tag{2.27}$$

$$= \left(v_{m,n} - \sqrt{\alpha_{m,n}}\mathbf{w}_m^H\mathbf{h}_{m,n}\mathbf{U}_{m,n}^{-1}\right)\mathbf{U}_{m,n}\left(v_{m,n} - \sqrt{\alpha_{m,n}}\mathbf{w}_m^H\mathbf{h}_{m,n}\mathbf{U}_{m,n}^{-1}\right)^H$$

$$+1 - \alpha_{m,n}\mathbf{w}_m^H\mathbf{h}_{m,n}\mathbf{U}_{m,n}^{-1}\mathbf{h}_{m,n}^H\mathbf{w}_m. \tag{2.28}$$

where $\mathbf{U}_{m,n} = \sum_{j=1}^{M}\sum_{i=1}^{N_j}\kappa_{j,i}^{m,n}\alpha_{j,i}\mathbf{h}_{m,n}^H\mathbf{w}_j\mathbf{w}_j^H\mathbf{h}_{m,n} + \sigma^2$, and

$$\kappa_{j,i}^{m,n} = \begin{cases} \eta_{m,n}, & \text{if}\quad j = m \quad\text{and}\quad i > n, \\ 1, & \text{otherwise.} \end{cases} \tag{2.29}$$

Check (2.28), the MSE is minimized only when $v_{m,n} = \sqrt{\alpha_{m,n}}\mathbf{w}_m^H\mathbf{h}_{m,n}\mathbf{U}_{m,n}^{-1}$. Under this condition, the minimum MSE is given by

$$e_{m,n} = 1 - \alpha_{m,n}\mathbf{w}_m^H\mathbf{h}_{m,n}\mathbf{U}_{m,n}^{-1}\mathbf{h}_{m,n}^H\mathbf{w}_m$$

$$= \frac{\mathbf{U}_{m,n} - \mathbf{h}_{m,n}^H\mathbf{w}_m\mathbf{w}_m^H\mathbf{h}_{m,n}\alpha_{m,n}}{\mathbf{U}_{m,n}}$$

$$= \frac{1}{1 + \gamma_{m,n}}, \tag{2.30}$$

and $v_{m,n}^{\text{MMSE}} = \sqrt{\alpha_{m,n}}\mathbf{w}_m^H\mathbf{h}_{m,n}\mathbf{U}_{m,n}^{-1}$ is the MMSE receiver.

Appendix B   Proof of Proposition 1

The dual function of the LP problem for the simplified design for multiple-antenna NOMA downlink for minimizing the total power consumption is given by

$$\mathcal{L}_2(\mathbf{P}) = \sum_{m=1}^{M} \sum_{n=1}^{N_m} P_{m,n} - \iota_{m,n} \left(|\mathbf{h}_{m,n}^H \mathbf{w}_m'|^2 P_{m,n} - \gamma_{m,n}^0 |\mathbf{h}_{m,n}^H \mathbf{w}_m'|^2 \sum_{i=1,i\neq n}^{N_m} \kappa_{m,i}^{m,n} P_{m,i} - \gamma_{m\ n}^0 \sigma^2 \right),$$
(2.31)

where $\iota_{m,n} \geq 0$ is the Lagrange multiplier related to the constraint condition in (2.23). According to the KKT conditions, we have

$$\frac{\partial \mathcal{L}_2(\mathbf{P})}{\partial P_{m,n}} = 1 - \iota_{m,n} |\mathbf{h}_{m,n}^H \mathbf{w}_m'|^2 + \sum_{i=1,i\neq n}^{N_m} \iota_{m,i} \gamma_{m,i}^0 |\mathbf{h}_{m,i}^H \mathbf{w}_m'|^2 \kappa_{m,n}^{m,i} = 0. \quad (2.32)$$

In other words, we have

$$\iota_{m,n} = \frac{1 + \sum_{i=1,i\neq n}^{N_m} \iota_{m,i} \gamma_{m,i}^0 |\mathbf{h}_{m,i}^H \mathbf{w}_m'|^2 \kappa_{m,n}^{m,i}}{|\mathbf{h}_{m,n}^H \mathbf{w}_m'|^2} > 0. \quad (2.33)$$

On the other hand, based on the KKT condition that $\iota_{m,n} \Big(|\mathbf{h}_{m,n}^H \mathbf{w}_m'|^2 F_{m,n} - \gamma_{m,n}^0 |\mathbf{h}_{m,n}^H \mathbf{w}_m'|^2 \sum_{i=1,i\neq n}^{N_m} \kappa_{m,i}^{m,n} P_{m,i} - \gamma_{m,n}^0 \sigma^2 \Big) = 0$, the following equality must hold true

$$|\mathbf{h}_{m,n}^H \mathbf{w}_m'|^2 P_{m,n} - \gamma_{m,n}^0 |\mathbf{h}_{m,n}^H \mathbf{w}_m'|^2 \sum_{i=1,i\neq n}^{N_m} \kappa_{m,i}^{m,n} P_{m,i} - \gamma_{m,n}^0 \sigma^2 = 0. \quad (2.34)$$

Rearranging (2.34), we have

$$\gamma_{m,n}' = \frac{|\mathbf{h}_{m,n}^H \mathbf{w}_m'|^2 P_{m,n}}{|\mathbf{h}_{m,n}^H \mathbf{w}_m'|^2 \sum_{i=1,i\neq n}^{N_m} \kappa_{m,i}^{m,n} P_{m,i} + \sigma^2} = \gamma_{m,n}^0. \quad (2.35)$$

Thus, we get the Proposition 1.

References

1. M.R. Palattella, M. Dohler, A. Grieco, G. Rizzo, J. Torsner, T. Engel, L. Ladid, Internet of things in the 5G era: enablers, architecture, and business models. IEEE J. Sel. Areas Commun. **34**(3), 510–527 (2016)
2. M. Shirvanimoghaddam, M. Dohler, S.J. Johnson, Massive non-orthogonal multiple access for cellular IoT: potentials and limitations. IEEE Commun. Mag. **55**(9), 55–61 (2017)
3. L. Dai, B. Wang, Y. Yuan, S. Han, I. Chih-lin, Z. Wang, Non-orthogonal multiple access for 5G: solutions, challenges, opportunities, and future research trends. IEEE Commun. Mag. **53**(9), 74–81 (2015)
4. M.S. Ali, H. Tabassum, E. Hossain, Dynamic user clustering and power allocation for uplink and downlink non-orthogonal multiple access (NOMA) systems. IEEE Access **4**, 6325–6343 (2016)
5. Y. Sun, D.W.K. Ng, Z. Ding, R. Schober, Optimal joint power and subcarrier allocation for full-duplex multicarrier non-orthogonal multiple access systems. IEEE Trans. Commun. **65**(3), 1077–1091 (2017)
6. H. Dai, A.F. Molisch, H.V. Poor, Downlink capacity of interference-limited MIMO systems with joint detection. IEEE Trans. Wirel. Commun. **3**(2), 442–453 (2004)
7. X. Chen, Z. Zhang, S. Chen, C. Wang, Adaptive mode selection for multiuser MIMO downlink employing rateless codes with QoS provisioning. IEEE Trans. Wirel. Commun. **11**(2), 790–799 (2012)
8. X. Chen, C. Yuen, Performance analysis and optimization for interference alignment over MIMO interference channels with limited feedback. IEEE Trans. Signal Process. **62**(7), 1785–1795 (2014)
9. Z. Ding, F. Adachi, H.V. Poor, The application of MIMO to non-orthogonal multiple access. IEEE Trans. Wireless Commun. **15**(1), 537–552 (2016)
10. X. Chen, Z. Zhang, C. Zhong, D.W.K. Ng, Exploiting multiple-antenna techniques for non-orthogonal multiple access. IEEE J. Sel. Areas Commun. **35**(10), 2207–2220 (2017)
11. Y. Liu, M. Elkashlan, Z. Ding, G.K. Karagiannidis, Fairness of user clustering in MIMO non-orthogonal multiple access systems. IEEE Commun. Lett. **20**(7), 1464–1468 (2016)
12. Z. Chen, Z. Ding, X. Dai, G.K. Karagiannidis, On the application of quasi-degradation to MISO-NOMA downlink. IEEE Trans. Signal Process. **64**(23), 6174–6189 (2016)
13. Z. Ding, R. Schober, H.V. Poor, A general MIMO framework for NOMA downlink and uplink transmission based on signal alignment. IEEE Trans. Wirel. Commun. **15**(6), 4438–4454 (2016)
14. W. Shin, M. Vaezi, B. Lee, D.J. Love, J. Lee, H.V. Poor, Coordinated beamforming for multi-cell MIMO-NOMA. IEEE Commun. Lett. **21**(1), 84–87 (2017)
15. Z. Chen, Z. Ding, X. Dai, Beamforming for combating inter-cluster and intra-cluster interference in hybrid NOMA systems. IEEE Access **4**, 4452–4463 (2016)
16. M.S. Ali, E. Hossain, D.I. Kim, Non-orthogonal multiple access (NOMA) for downlink multiuser MIMO systems: user clustering, beamforming, and power allocation. IEEE Access **5**, 565–577 (2017)
17. X. Sun, D. Duran-Herrmann, Z. Zhong, Y. Yang, Non-orthogonal multiple access with weighted sum-rate optimization for downlink broadcast channel, in *Proceedings of IEEE MILCOM*, Oct 2015, pp. 1176–1181
18. J. Kim, J. Koh, J. Kang, K. Lee, J. Kang, Design of user clustering and precoding for downlink non-orthogonal multiple access (NOMA), in *Proceedings of IEEE MILCOM*, Oct 2015, pp. 1170–1175
19. M.F. Hanif, Z. Ding, T. Ratnarajah, G.K. Karagiannidis, A minorization-maximization method for optimizing sum rate in the downlink of non-orthogonal multiple access systems. IEEE Trans. Signal Process. **64**(1), 76–88 (2016)
20. C. Zhong, Z. Zhang, Non-orthogonal multiple access with cooperative full-duplex relaying. IEEE Commun. Lett. **20**(12), 2478–2481 (2016)

21. P. Li, R.C. de Lamare, R. Fa, Multiple feedback successive interference cancellation detection for multiuser MIMO systems. IEEE Trans. Wirel. Commun. **10**(8), 2434–2439 (2011)

22. S.M.R. Islam, N. Avazov, O.A. Dobre, K.-S. Kwak, Power-domain non-orthogonal multiple access (NOMA) in 5G: potentials and challenges. IEEE Commun. Surv. Tuts **19**(2), 721–742 (2017)

23. H. Sun, B. Xie, R.Q. Hu, G. Wu, Non-orthogonal multiple access with SIC error propagation in downlink wireless MIMO networks, in *Proceedings of IEEE VTC 2016-Fall*, Invited paper, Sept 2016, pp. 1–5

24. X. Chen, Z. Zhang, C. Zhong, R. Jia, D.W.K. Ng, Fully non-orthogonal communication for massive access. IEEE Trans. Commun. **66**(4), 1717–1731 (2018)

25. S.S. Christensen, R. Agarwal, E. de Carvalho, J.M. Cioffi, Weighted sum-rate maximization using weighted MMSE for MIMO-BC beamforming design. IEEE Trans. Wirel. Commun. **7**(12), 4792–4799 (2008)

26. Q. Shi, M. Razaviyayn, Z.-Q. Luo, C. He, An iteratively weighted MMSE approach to distributed sum-utility maximization for a MIMO interfering broadcast channel. IEEE Trans. Signal Process. **59**(9), 4331–4340 (2011)

27. B.S. Krongold, K. Ramchandran, D.L. Jones, Computationally efficient optimal power allocation algorithms for multicarrier communication systems. IEEE Trans. Commun. **48**(1), 23–27 (2000)

28. S. Boyd, L. Vandenberghe, *Convex Optimization* (Cambridge University Press, Cambridge, 2004)

29. M. Grant, S. Boyd, CVX: Matlab software for disciplined convex programming. [Online]: http://cvxr.com/cvx

Chapter 3
Massive Access with Channel Quantization Codebook

Abstract In this chapter, we provide a comprehensive solution for the design, analysis, and optimization of a cellular IoT operated in FDD mode. First, we design a massive access framework based on channel quantization codebook. Then, we analyze the performance of the considered system, and derive exact closed-form expressions for average transmission rates in terms of transmit power, CSI accuracy, transmission mode, and channel conditions. For further enhancing the system performance, we optimize three key parameters, i.e., transmit power, feedback bits, and transmission mode. Especially, we propose a low-complexity joint optimization scheme, so as to fully exploit the potential of multiple-antenna techniques for massive access. Moreover, through asymptotic analysis, we reveal the impact of system parameters on average transmission rates, and hence present some guidelines on the design of massive access systems. Finally, simulation results validate our theoretical analysis, and show that a substantial performance gain can be obtained over traditional orthogonal multiple access (OMA) techniques in the scenario of massive access for the cellular IoT.

3.1 Introduction

Non-orthogonal multiple access (NOMA) has been widely recognized as an enabling technique for supporting massive access in the 5G cellular IoT [1–3]. However, the severe co-channel interference caused by NOMA degrades the performance of the cellular IoT, and thus it is difficult to satisfy the quality of service (QoS) requirements. As is well known that the multiple-antenna technology is a powerful interference mitigation scheme [4, 5], hence, can be naturally applied to NOMA systems for achieving massive access [6, 7]. In [8], the authors proposed a beamforming scheme for combating inter-cluster and intra-cluster interference in a NOMA downlink, where the base station (BS) was equipped with multiple antennas and the user equipments (UEs) have a single antenna each. A more general setup was considered in [9], where both the BS and the UEs are multiple-antenna devices. By exploiting multiple antennas at the BS and the UEs, a signal alignment scheme was proposed to mitigate both the intra-cluster and inter-cluster interference. It is worth pointing out that the implementation of the two above schemes requires full channel state information (CSI) at the BS, which is usually difficult and costly in practice. To circumvent the difficulty in CSI acquisition, random beamforming was adopted in [10], which inevitably leads to performance loss. Alternatively, the work in [11] suggested to employ zero-forcing (ZF) detection at the multiple-antenna UEs for inter-cluster interference cancelation. However, the ZF scheme requires that the number of antennas at each UE is greater than the number of antennas at the BS, which is in general impractical.

To effectively realize the potential benefits of multiple-antenna techniques for massive access, the amount and quality of CSI available at the BS plays a key role [12]. In practice, the CSI can be obtained in several different ways [13, 14]. For frequency duplex division (FDD) systems, the downlink CSI is usually first estimated and quantized at the UEs, and then is conveyed back to the BS via a feedback link [15]. In other words, the BS in the practical cellular IoT has access to only partial CSI. As a result, there will be residual inter-cluster and intra-cluster interference. To the best of the authors' knowledge, previous works only consider two extreme cases with full CSI or no CSI, the design, analysis and optimization of FDD-based massive NOMA systems with partial CSI remains an uncharted area.

Motivated by this, we present a comprehensive study on the impact of partial CSI on the design, analysis, and optimization of massive NOMA downlink systems for the cellular IoT. Specifically, we consider heterogeneous downlink channels, and the BS equipped with arbitrarily multiple antennas has different CSI accuracies about the downlink channels. The major contributions of this chapter are summarized as follows:

1. We design a general framework for massive NOMA downlink communications in the FDD-based cellular IoT, including user clustering, CSI estimation, superposition coding, transmit beamforming, and SIC.
2. We analyze the performance of the proposed massive NOMA, and derive exact expressions for the average transmission rates of each UE in an arbitrary cluster. The average transmission rate is a function of transmit power, CSI accuracy, transmission mode, and channel conditions.
3. We optimize three key parameters of massive NOMA, namely, transmit power, feedback bits, and transmission mode. In particular, we present closed-form expressions for the power allocation and feedback distribution. For mode selection, we show that the mode of two UEs in a cluster is optimal in practical cases with moderate and high CSI accuracy, which provides theoretical justification for the two-user setup in the previous works [6–11, 16, 17]. Finally, a low complexity joint optimization scheme of transmit power, feedback bits, and transmission mode is proposed.
4. Through asymptotic analysis of average transmission rates, several key insights are obtained.

   (a) Imperfect CSI results in residual inter-cluster interference at UEs. Thus, there exists a performance gap between practical NOMA with imperfect CSI and ideal NOMA with perfect CSI. The performance gap is an increasing function of transmit power of information signal and a decreasing function of CSI accuracy. In order to maintain a constant performance, spatial resolution in CSI feedback should be increased as transmit power of information signal grows.

(b) From the perspective of maximizing the sum rate, arranging all UEs in one cluster is optimal if there is no CSI at the BS, while the best option is to arrange one UE in each cluster if there is perfect CSI at the BS.

(c) In the interference-limited scenario, the average transmission rate for the 1st UE with the strongest channel gain in each cluster increases linearly proportionally to the number of feedback bits. Under the noise-limited condition, the average transmission rate is independent of CSI accuracy.

(d) In the interference-limited case, equal power allocation among all UEs asymptotically approaches the optimal performance.

The rest of this chapter is organized as follows: Sect. 3.2 gives a brief introduction of the considered FDD-based cellular IoT, and designs a massive NOMA framework. Section 3.3 first analyzes the average transmission rates in presence of imperfect CSI, and then proposes three performance optimization schemes. Section 3.4 derives the average transmission rates in two extreme cases through asymptotic analysis, and presents some system design guidelines. Section 3.5 provides simulation results to validate the effectiveness of the proposed schemes. Finally, Sect. 3.6 concludes the chapter.

Notations We use bold upper (lower) letters to denote matrices (column vectors), $(\cdot)^H$ to denote conjugate transpose, $E[\cdot]$ to denote expectation, $\| \cdot \|$ to denote the L_2-norm of a vector, $| \cdot |$ to denote the absolute value, $\overset{d}{=}$ to denote the equality in distribution, $\lfloor x \rfloor$ to denote the maximum integer not larger than x, and $\mathscr{C}$ to denote the set of complex number. The acronym i.i.d. means "independent and identically distributed", pdf means "probability density function", and cdf means "cumulative distribution function".

3.2 System Model and Problem Formulation

We consider a cellular IoT, where there exist some devices with low-speed mobility. A BS with N_t antennas broadcasts messages to K single antenna UEs, cf. Fig. 3.1. The UEs in the same direction but with distinctive propagation distances are grouped into a cluster, and the UEs in a cluster share a transmit beam. Without loss of generality, we also assume that the K UEs are partitioned into M clusters, and the mth cluster contains N_m UEs. To facilitate the following presentation, we use $\alpha_{n,k}^{1/2}\mathbf{h}_{n,k}$ to denote the M-dimensional channel vector from the BS to the kth UE in the nth cluster, where $\alpha_{n,k}$ is the large-scale channel fading, and $\mathbf{h}_{n,k}$ is the small-scale channel fading following zero mean complex Gaussian distribution with unit

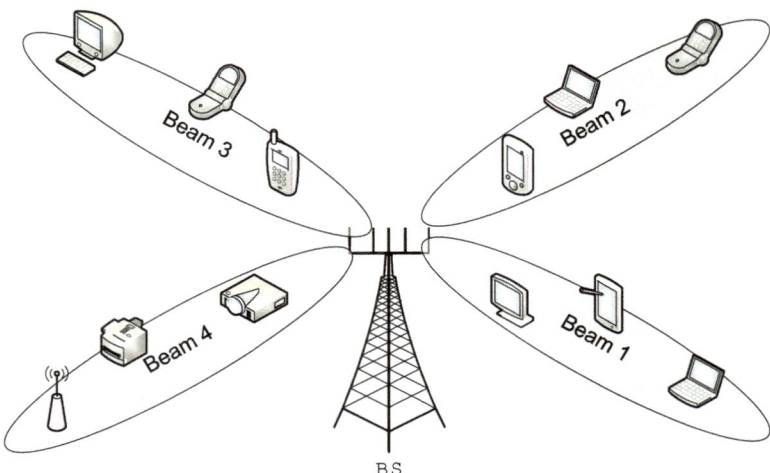

Fig. 3.1 A massive access model of the cellular IoT in FDD mode

variance. It is assumed that $\alpha_{n,k}$ remains constant for a relatively long period, while $\mathbf{h}_{n,k}$ keeps unchanged in a time slot but varies independently over time slots.

Due to the low-speed mobility of the UEs, it is difficult for the BS to obtain full CSI. We assume that the cellular IoT operated in FDD mode, and the CSI is conveyed from the UEs to the BS through a feedback link. Since the feedback link is rate-constrained, CSI at the UEs should first be quantized. Specifically, the $\text{UE}_{m,n}$ chooses an optimal codeword from a predetermined quantization codebook $\mathscr{B}_{m,n} = \{\tilde{\mathbf{h}}_{m,n}^{(1)}, \ldots, \tilde{\mathbf{h}}_{m,n}^{(2^{B_{m,n}})}\}$ of size $2^{B_{m,n}}$, where $\tilde{\mathbf{h}}_{m,n}^{(j)}$ is the jth codeword of unit norm and $B_{m,n}$ is the number of feedback bits. Mathematically, the codeword selection criterion is given by

$$j^{\star} = \arg \max_{1 \le j \le 2^{B_{m,n}}} \left| \mathbf{h}_{m,n}^{H} \tilde{\mathbf{h}}_{m,n}^{(j)} \right|^{2}. \tag{3.1}$$

Then, the $\text{UE}_{m,n}$ conveys the index $j^{\star}$ to the BS with $B_{m,n}$ feedback bits, and the BS recoveries the quantized CSI $\tilde{\mathbf{h}}_{m,n}^{(j^{\star})}$ from the same codebook. In other words, the BS only gets the phase information by using the feedback scheme based on a quantization codebook. However, as shown in below, the phase information is sufficient for the design of spatial beamforming. Similarly, the relation between the real CSI and the obtained CSI in FDD mode can be approximated as [18, 19]

$$\tilde{\mathbf{h}}_{m,n} = \sqrt{\varrho_{m,n}} \tilde{\mathbf{h}}_{m,n}^{\star} + \sqrt{1 - \varrho_{m,n}} \tilde{\mathbf{e}}_{m,n}, \tag{3.2}$$

where $\tilde{\mathbf{h}}_{m,n} = \frac{\mathbf{h}_{m,n}}{\|\mathbf{h}_{m,n}\|}$ is the phase of the channel $\mathbf{h}_{m,n}$, $\tilde{\mathbf{h}}_{m,n}^{\star}$ is the quantized phase information, $\tilde{\mathbf{e}}_{m,n}$ is the quantization error vector with uniform distribution, and

$\varrho_{m,n} = 1 - 2^{-\frac{B_{m,n}}{N_t - 1}}$ is the associated correlation coefficient or CSI accuracy. Thus, it is possible to improve the CSI accuracy by increasing the size of quantization codebook for a given number of antennas N_t at the BS.

Based on the available CSI, the BS constructs one transmit beam for each cluster, so as to mitigate or even completely cancel the inter-cluster interference. To strike balance between system performance and implementation complexity, we adopt zero-force beamforming (ZFBF) at the BS. We take the deign of beam $\mathbf{w}_i$ for the ith cluster as an example. First, we construct a complementary matrix $\bar{\mathbf{H}}_i$ as:

$$\bar{\mathbf{H}}_i = [\tilde{\mathbf{h}}_{1,1}^{\star}, \cdots, \tilde{\mathbf{h}}_{1,N_1}^{\star}, \cdots, \tilde{\mathbf{h}}_{i-1,1}^{\star}, \cdots, \tilde{\mathbf{h}}_{i-1,N_{i-1}}^{\star}, \tilde{\mathbf{h}}_{i+1,1}^{\star}, \cdots, \tilde{\mathbf{h}}_{M,N_M}^{\star}]^H. \tag{3.3}$$

Then, we perform singular value decomposition (SVD) on $\bar{\mathbf{H}}_i$ and obtain its right singular vectors $\mathbf{u}_{i,j}$, $j = 1, \cdots, N_u$, with respect to the zero singular values, where N_u is the number of zero singular values. Finally, we can design the beam as $\mathbf{w}_i = \sum_{j=1}^{N_u} \theta_{i,j} \mathbf{u}_{i,j}$, where $\theta_{i,j} > 0$ is a weight such that $\sum_{j=1}^{N_u} \theta_{i,j} = 1$. Thus, the received signal at the UE$_{m,n}$ is given by

$$y_{m,n} = \sqrt{\alpha_{m,n}} \mathbf{h}_{m,n}^H \sum_{i=1}^{M} \mathbf{w}_i s_i + n_{m,n}$$

$$= \sqrt{\alpha_{m,n}} \mathbf{h}_{m,n}^H \mathbf{w}_m s_m + \sqrt{\alpha_{m,n}(1 - \rho_{m,n})} \mathbf{e}_{m,n}^H \sum_{i=1, i \neq m}^{M} \mathbf{w}_i s_i + n_{m,n}, \tag{3.4}$$

where $s_i = \sum_{j=1}^{N_j} \sqrt{P_{i,j}} s_{i,j}$ is the superposition coded signal with $P_{i,j}$ and $s_{i,j}$ being transmit power and transmit signal for the UE$_{i,j}$, and $n_{m,n}$ is the AWGN with unit variance. In general, $P_{i,j}^S$ should be carefully allocated to distinguish the UEs in the power domain, which we will discuss in detail below. Note that Eq. (3.4) holds true due to the fact that $\mathbf{h}_{m,n}^H \mathbf{w}_i = \sqrt{\varrho_{m,n}} \|\mathbf{h}_{m,n}\| (\tilde{\mathbf{h}}_{m,n}^{\star})^H \mathbf{w}_i + \sqrt{1 - \varrho_{m,n}} \|\mathbf{h}_{m,n}\| \tilde{\mathbf{e}}_{m,n}^H \mathbf{w}_i = \sqrt{1 - \varrho_{m,n}} \|\mathbf{h}_{n,k}\| \tilde{\mathbf{e}}_{m,n}^H \mathbf{w}_i \overset{d}{=} \sqrt{1 - \varrho_{m,n}} \mathbf{e}_{m,n}^H \mathbf{w}_i$, where $\overset{d}{=}$ denotes the equality in distribution, and $\mathbf{e}_{n,k}$ is a random vector with i.i.d. zero mean and unit variance complex Gaussian distributed entries. With perfect CSI at the BS, i.e., $\rho_{m,n} = 1$, the inter-cluster interference can be completely cancelled.

Although ZFBF at the BS can mitigate partial inter-cluster interference from the other clusters, there still exists intra-cluster interference from the same cluster. In order to improve the received signal quality, the UE conducts SIC. Without loss of generality, we assume that the effective channel gains in the ith cluster have the following order:

$$|\sqrt{\alpha_{i,1}} \mathbf{h}_{i,1}^H \mathbf{w}_i|^2 \geq \cdots \geq |\sqrt{\alpha_{i,N_i}} \mathbf{h}_{i,N_i}^H \mathbf{w}_i|^2. \tag{3.5}$$

It is reasonably assumed that the BS may know the UEs' effective gains through the channel quality indicator (CQI) messages, and then determines the user order in (3.5). Thus, in the ith cluster, the jth UE can always successfully decode the lth UE's signal, $\forall l > j$, if the lth UE can decode its own signal. As a result, the jth UE can subtract the interference from the lth UE in the received signal before decoding its own signal. After SIC, the signal-to-interference-plus-noise ratio (SINR) at the $\text{UE}_{m,n}$ is given by

$$\gamma_{m,n} = \frac{\alpha_{m,n} |\mathbf{h}_{m,n}^H \mathbf{w}_m|^2 P_{m,n}}{\underbrace{\alpha_{m,n} |\mathbf{h}_{m,n}^H \mathbf{w}_m|^2 \sum_{j=1}^{n-1} P_{m,j}}_{\text{Intra-cluster interference}} + \underbrace{\alpha_{m,n}(1 - \rho_{m,n}) \sum_{i=1,i\neq m}^{M} |\mathbf{e}_{m,n}^H \mathbf{w}_i|^2 \sum_{l=1}^{N_i} P_{i,l}}_{\text{Inter-cluster interference}} + \underbrace{1}_{\text{AWGN}}},$$

(3.6)

where the first term in the denominator of (3.6) is the residual intra-cluster interference after SIC at the UE, the second one is the residual inter-cluster interference after ZFBF at the BS, and the third one is the AWGN. For the 1st UE in each cluster, there is no intra-cluster interference, since it can completely eliminate the intra-cluster interference. Note that in this chapter, we assume that perfect SIC can be performed at the UEs. In practical NOMA systems, SIC might be imperfect due to a limited computational capability at the UEs. Thus, there exists residual intra-cluster interference from the weaker UEs even after SIC [20]. As discussed in the last chapter, the residual intra-interference due to imperfect SIC can be modeled as a linear function of the interfering signal. Thus, the residual intra-interference caused by imperfect SIC would not change the structure of the SINR. In other words, the analysis and optimization based on (3.6) can be extended to the case of imperfect SIC directly.

3.3   Performance Analysis and Optimization

In this section, we concentrate on performance analysis and optimization of a massive access system with imperfect CSI based on channel quantization codebook. Specifically, we first derive closed-form expressions for the average transmission rates of the 1st UE and the other UEs, and then propose separate and joint optimization schemes of transmit power, feedback bits, and transmit mode, so as to maximize the average sum rate of the system.

3.3.1   Average Transmission Rate

We start by analyzing the average transmission rate of the $\text{UE}_{m,n}$. First, we consider the case $n > 1$. According to the definition, the corresponding average transmission rate can be computed as

$$R_{m,n} = \mathrm{E}\left[\log_2\left(1 + \gamma_{m,n}\right)\right]$$

$$= \mathrm{E}\left[\log_2\left(\frac{\alpha_{m,n}|\mathbf{h}_{m,n}^H\mathbf{w}_m|^2\sum_{j=1}^{n}P_{m,j} + \alpha_{m,n}(1-\rho_{m,n})\sum_{i=1,i\neq m}^{M}|\mathbf{e}_{m,n}^H\mathbf{w}_i|^2\sum_{l=1}^{N_i}P_{i,l} + 1}{\alpha_{m,n}|\mathbf{h}_{m,n}^H\mathbf{w}_m|^2\sum_{j=1}^{n-1}P_{m,j} + \alpha_{m,n}(1-\rho_{m,n})\sum_{i=1,i\neq m}^{M}|\mathbf{e}_{m,n}^H\mathbf{w}_i|^2\sum_{l=1}^{N_i}P_{i,l} + 1}\right)\right]$$

$$= \mathrm{E}\left[\log_2\left(\alpha_{m,n}|\mathbf{h}_{m,n}^H\mathbf{w}_m|^2\sum_{j=1}^{n}P_{m,j} + \alpha_{m,n}(1-\rho_{m,n})\sum_{i=1,i\neq m}^{M}|\mathbf{e}_{m,n}^H\mathbf{w}_i|^2\sum_{l=1}^{N_i}P_{i,l} + 1\right)\right]$$

$$-\mathrm{E}\left[\log_2\left(\alpha_{m,n}|\mathbf{h}_{m,n}^H\mathbf{w}_m|^2\sum_{j=1}^{n-1}P_{m,j} + \alpha_{m,n}(1-\rho_{m,n})\sum_{i=1,i\neq m}^{M}|\mathbf{e}_{m,n}^H\mathbf{w}_i|^2\sum_{l=1}^{N_i}P_{i,l} + 1\right)\right].$$

$$(3.7)$$

Note that the average transmission rate in (3.7) can be expressed as the difference of two terms, which have a similar form. Hence, we concentrate on the derivation of the first term. For notational convenience, we use W to denote the term $\alpha_{m,n}|\mathbf{h}_{m,n}^H\mathbf{w}_m|^2\sum_{j=1}^{n}P_{m,j} + \alpha_{m,n}(1-\rho_{m,n})\sum_{i=1,i\neq m}^{M}|\mathbf{e}_{m,n}^H\mathbf{w}_i|^2\sum_{l=1}^{N_i}P_{i,l}$. To compute the first expectation, the key is to obtain the probability density function (pdf) of W. Checking the first random variable $|\mathbf{h}_{m,n}^H\mathbf{w}_m|^2$ in W, since $\mathbf{w}_m$ of unit norm is designed independent of $\mathbf{h}_{m,n}$, $|\mathbf{h}_{m,n}^H\mathbf{w}_m|^2$ is χ^2 distributed with 2 degrees of freedom [21]. Similarly, $|\mathbf{e}_{m,n}^H\mathbf{w}_i|^2$ also has the distribution $\chi^2(2)$. Therefore, W can be considered as a weighted sum of N random variables with $\chi^2(2)$ distribution. According to [22], W is a nested finite weighted sum of M Erlang pdfs, whose pdf is given by

$$f_W(x) = \sum_{i=1}^{M} \Xi_M\left(i, \{\eta_{m,n}^q\}_{q=1}^M\right) g(x, \eta_{m,n}^i), \qquad (3.8)$$

where

$$\eta_{m,n}^q = \begin{cases} \alpha_{m,n}\sum_{j=1}^{n}P_{q,j} & \text{if } q = m \\ \alpha_{m,n}(1-\rho_{m,n})\sum_{l=1}^{N_q}P_{q,l} & \text{if } q \neq m \end{cases},$$

$$g(x, \eta_{m,n}^i) = \frac{1}{\eta_{m,n}^i}\exp\left(-\frac{x}{\eta_{m,n}^i}\right),$$

$$\Xi_N\left(i, \{\eta_{m,n}^q\}_{q=1}^M\right) = \frac{(-1)^{M-1}\eta_{m,n}^i}{\prod\limits_{l=1}^M \eta_{m,n}^l}\prod_{s=1}^{M-1}\left(\frac{1}{\eta_{m,n}^i}-\frac{1}{\eta_{m,n}^{s+U(s-i)}}\right)^{-1},$$

and $U(x)$ is the well-known unit step function defined as $U(x \geq 0) = 1$ and zero otherwise. It is worth pointing out that the weights Ξ_N are constant for given $\{\eta_{m,n}^q\}_{q=1}^M$. Hence, the first expectation in (3.7) can be computed as

$$\begin{aligned}
E[\log_2(1+W)] &= \int_0^\infty \log_2(1+x)f_W(x)dx \\
&= \sum_{i=1}^M \Xi_M\left(i, \{\eta_{m,n}^q\}_{q=1}^M\right)\int_0^\infty \log_2(1+x)\frac{1}{\eta_{m,n}^i}\exp\left(-\frac{x}{\eta_{m,n}^i}\right)dx \\
&= -\frac{1}{\ln(2)}\sum_{i=1}^M \Xi_M\left(i, \{\eta_{m,n}^q\}_{q=1}^M\right)\exp\left(\frac{1}{\eta_{m,n}^i}\right)E_i\left(-\frac{1}{\eta_{m,n}^i}\right),
\end{aligned}$$

(3.9)

where $E_i(x) = \int_{-\infty}^x \frac{\exp(t)}{t}dt$ is the exponential integral function. Equation (3.9) follows from [23, Eq. (4.3372)]. Similarly, we use V to denote $\alpha_{m,n}|h_{m,n}^H w_m|^2\sum_{j=1}^{n-1}P_{m,j}+\alpha_{m,n}(1-\rho_{m,n})\sum_{i=1,i\neq m}^M |e_{m,n}^H w_i|^2\sum_{l=1}^{N_i}P_{i,l}$ in the second term of (3.7). Thus, the second expectation term can be computed as

$$E[\log_2(1+V)] = -\frac{1}{\ln(2)}\sum_{i=1}^M \Xi_M\left(i, \{\beta_{m,n}^v\}_{v=1}^M\right)\exp\left(\frac{1}{\beta_{m,n}^i}\right)E_i\left(-\frac{1}{\beta_{m,n}^i}\right),$$

(3.10)

where

$$\beta_{m,n}^v = \begin{cases}\alpha_{m,n}\sum\limits_{j=1}^{n-1}P_{v,j} & \text{if } v=m \\ \alpha_{m,n}(1-\rho_{m,n})\sum\limits_{l=1}^{N_v}P_{v,l} & \text{if } v\neq m\end{cases}.$$

Hence, we can obtain the average transmission rate for the $UE_{m,n}$ as follows

$$\begin{aligned}
R_{m,n} = &\frac{1}{\ln(2)}\sum_{i=1}^M \Xi_M\left(i, \{\beta_{m,n}^v\}_{v=1}^M\right)\exp\left(\frac{1}{\beta_{m,n}^i}\right)E_i\left(-\frac{1}{\beta_{m,n}^i}\right) \\
&-\frac{1}{\ln(2)}\sum_{i=1}^M \Xi_N\left(i, \{\eta_{m,n}^q\}_{q=1}^M\right)\exp\left(\frac{1}{\eta_{m,n}^i}\right)E_i\left(-\frac{1}{\eta_{m,n}^i}\right). \quad (3.11)
\end{aligned}$$

Then, we consider the case $n = 1$. Since the first UE can decode all the other UEs' signals in the same cluster, there is no intra-cluster interference. In this case, the corresponding average transmission rate reduces to

$$R_{m,1} = \frac{1}{\ln(2)} \sum_{i=1}^{M-1} \varXi_{M-1} \left(i, \{\beta_{m,1}^v\}_{v=1}^{M-1} \right) \exp \left(\frac{1}{\beta_{m,1}^i} \right) E_i \left(-\frac{1}{\beta_{m,1}^i} \right)$$

$$-\frac{1}{\ln(2)} \sum_{i=1}^{M} \varXi_N \left(i, \{\eta_{m,1}^q\}_{q=1}^{M} \right) \exp \left(\frac{1}{\eta_{m,1}^i} \right) E_i \left(-\frac{1}{\eta_{m,1}^i} \right), \quad (3.12)$$

where

$$\eta_{m,1}^q = \begin{cases} \alpha_{m,1} P_{q,1} & \text{if } q = m \\ \alpha_{m,1}(1 - \rho_{m,1}) \sum_{l=1}^{N_q} P_{q,l} & \text{if } q \neq m \end{cases},$$

and

$$\beta_{m,1}^v = \begin{cases} \alpha_{m,1}(1 - \rho_{m,1}) \sum_{l=1}^{N_v} P_{v,l} & \text{if } v < m \\ \alpha_{m,1}(1 - \rho_{m,1}) \sum_{l=1}^{N_{v+1}} P_{v+1,l} & \text{if } v \geq m \end{cases}.$$

Combing (3.11) and (3.12), it is easy to evaluate the performance of a massive access system with arbitrary system parameters and channel conditions. In particular, it is possible to reveal the impact of system parameters, i.e., transmit power, CSI accuracy, and transmission mode.

3.3.2   Power Allocation

From (3.11) and (3.12), it is easy to observe that with imperfect CSI, transmit power has a great impact on average transmission rates. On one hand, increasing the transmit power can enhance the desired signal strength. On the other hand, it also increases the interference. Thus, it is desired to distribute the transmit power according to channel conditions.

To maximize the sum rate of the considered massive access system subject to a total power constraint, we have the following optimization problem:

$$J_1 : \max_{P_{m,n}} \sum_{m=1}^{M} \sum_{n=1}^{N_m} R_{m,n} \quad (3.13)$$

$$\text{s.t.C1}: \sum_{m=1}^{M} \sum_{n=1}^{N_m} P_{m,n} \leq P_{tot}$$

$$\text{C2}: P_{m,n} > 0,$$

where P_{tot} is the maximum total transmit power budget. It is worth pointing out that in certain scenarios, user fairness might be of particular importance. To guarantee user fairness, one can replace the objective function of J_1 with the maximization of a weighted sum rate, where the weights can directly affect the power allocation and thus the UEs' rates. Unfortunately, J_1 is not a convex problem due to the complicated expression for the objective function. Thus, it is difficult to directly provide a closed-form solution for the optimal transmit power. As a compromise solution, we propose an effective power allocation scheme based on the following important observation of the massive access system:

Lemma 1 *The inter-cluster interference is dependent of power allocation between the clusters, while the intra-cluster interference is determined by power allocation among the UEs in the same cluster.*

Proof A close observation of the inter-cluster interference $\alpha_{m,n}(1 - \rho_{m,n}) \sum_{i=1,i\neq m}^{M} |\mathbf{e}_{m,n}^{H} \mathbf{w}_i|^2 \sum_{l=1}^{N_i} P_{i,l}$ in (3.6) indicates that $\sum_{l=1}^{N_i} P_{i,l}$ is the total transmit power for the ith cluster, which suggests that inter-cluster power allocation does not affect the inter-cluster interference. $\square$

Inspired by Lemma 1, the power allocation scheme can be divided into two steps. In the first step, the BS distributes the total power among the M clusters. In the second step, each cluster individually carries out power allocation subject to the power constraint determined by the first step. In the following, we give the details of the two-step power allocation scheme. First, we design the power allocation between the clusters from the perspective of minimizing inter-cluster interference. For the ith cluster, the average aggregate interference to the other clusters is given by

$$I_i = E\left[\sum_{m=1,m\neq i}^{M} \sum_{n=1}^{N_m} \alpha_{m,n}(1 - \rho_{m,n})|\mathbf{e}_{m,n}^{H} \mathbf{w}_i|^2 \sum_{l=1}^{N_i} P_{i,l} \right]$$

$$= \left(\sum_{m=1,m\neq i}^{M} \sum_{n=1}^{N_m} \alpha_{m,n}(1 - \rho_{m,n}) \right) P_i, \tag{3.14}$$

where $P_i = \sum_{l=1}^{N_i} P_{i,l}$ is the total transmit power of the ith cluster. Equation (3.14) follows the fact that $E[|\mathbf{e}_{m,n}^{H} \mathbf{w}_i|^2] = 1$. Intuitively, a large interference coefficient

$\sum_{m=1,m\neq i}^{M} \sum_{n=1}^{N_m} \alpha_{m,n}(1 - \rho_{m,n})$ means a more severe inter-cluster interference caused by the ith cluster. In order to mitigate the inter-cluster interference for improving the average sum rate, we propose to distribute the power proportionally to the reciprocal of interference coefficient. Specifically, the transmit power for the ith cluster can be computed as

$$P_i = \frac{\left(\sum_{m=1,m\neq i}^{M} \sum_{n=1}^{N_m} \alpha_{m,n}(1 - \rho_{m,n})\right)^{-1}}{\sum_{l=1}^{M} \left(\sum_{m=1,m\neq l}^{M} \sum_{n=1}^{N_m} \alpha_{m,n}(1 - \rho_{m,n})\right)^{-1}} P_{tol}. \tag{3.15}$$

Then, we allocate the power in the cluster for further increasing the average sum rate. According to the nature of NOMA techniques, the first UE not only has the strongest effective channel gain for the desired signal, but also generates a weak interference to the other UEs. On the contrary, the last UE has the weakest effective channel gain for the desired signal, and also produces a strong interference to the other UEs. Thus, from the perspective of maximizing the sum of average rate, it is better to allocate the power based on the following criterion:

$$P_{m,1} \geq \cdots \geq P_{m,n} \geq \cdots \geq P_{m,N_m}. \tag{3.16}$$

On the other hand, in order to facilitate SIC, the NOMA in general requires the transmit powers in a cluster to follow a criterion below [11]:

$$P_{m,1} \leq \cdots \leq P_{m,n} \leq \cdots \leq P_{m,N_m}. \tag{3.17}$$

Under this condition, the UE performs SIC according to the descending order of the user index, namely the ascending order of the effective channel gain. Specifically, the nth UE cancels the interference from the N_mth to the $(n + 1)$th UE in sequence. Thus, the SINR for decoding each interference signal is the highest, which facilitates SIC at UEs [24].

To simultaneously fulfill the above two criterions, we propose to equally distribute the powers within a cluster, namely

$$P_{m,n} = P_m/N_m. \tag{3.18}$$

Substituting (3.15) into (3.18), the transmit power for the UE$_{m,n}$ can be computed as

$$P_{m,n} = \frac{\left(\sum_{i=1,i\neq m}^{M} \sum_{j=1}^{N_i} \alpha_{i,j}(1 - \rho_{i,j})\right)^{-1}}{N_m \left(\sum_{l=1}^{M} \left(\sum_{i=1,i\neq l}^{M} \sum_{j=1}^{N_i} \alpha_{i,j}(1 - \rho_{i,j})\right)^{-1}\right)} P_{tol}. \tag{3.19}$$

Thus, we can distribute the transmit power based on (3.19) for given channel statistical information and the CSI accuracy, which has a quite low computational complexity.

Remarks We note that path loss coefficients $\alpha_{m,n}$, $\forall m, n$, remain constant for a relatively long time, and it is easy to obtain at the BS via long-term measurement. Hence, the proposed power allocation scheme incurs a low system overhead, and can be implemented with low complexity.

3.3.3 Feedback Distribution

As discussed earlier, the accuracy of quantized CSI relies on the size of codebook $2^{B_{m,n}}$, where $B_{m,n}$ is the number of feedback bits from the $\mathrm{UE}_{m,n}$. As observed in (3.11) and (3.12), it is possible to decrease the interference by increasing feedback bits. However, due to the rate constraint on the feedback link, the total number of feedback bits is limited. Therefore, it is of great importance to optimize the feedback bits among the UEs for performance enhancement.

According to the received SNR in (3.6), the CSI accuracy only affects the inter-cluster interference. Thus, it makes sense to optimize the feedback bits to minimizing the average sum of inter-cluster interference given by

$$
\begin{aligned}
I_{\text{inter}} &= \mathrm{E}\left[\sum_{m=1}^{M} \sum_{n=1}^{N_m} \alpha_{m,n}(1 - \varrho_{m,n}) \sum_{i=1,i\neq m}^{M} |\mathbf{e}_{m,n}^{H}\mathbf{w}_i|^2 \sum_{l=1}^{N_i} P_{i,l} \right] \\
&= \sum_{m=1}^{M} \sum_{n=1}^{N_m} \alpha_{m,n} \sum_{i=1,i\neq m}^{M} P_i 2^{-\frac{B_{m,n}}{N_t-1}}.
\end{aligned}
\tag{3.20}
$$

Hence, the optimization problem for feedback bits distribution can be expressed as

$$
J_2 : \min_{B_{m,n}} \sum_{m=1}^{M} \sum_{n=1}^{N_m} \alpha_{m,n} \sum_{i=1,i\neq m}^{M} P_i 2^{-\frac{B_{m,n}}{N_t-1}}
\tag{3.21}
$$

$$
\text{s.t.C3} : \sum_{m=1}^{M} \sum_{n=1}^{N_m} B_{m,n} \leq B_{\text{tot}},
$$

$$
\text{C4} : B_{m,n} \geq 0,
$$

where B_{tot} is an upper bound on the total number of feedback bits. J_2 is an integer programming problem, hence is difficult to solve. To tackle this challenge, we relax the integer constraint on $B_{m,n}$. In this case, according to the fact that

$$\sum_{m=1}^{M}\sum_{n=1}^{N_m}\alpha_{m,n}\sum_{i=1,i\neq m}^{M}P_i 2^{-\frac{B_{m,n}}{N_t-1}} \geq K\left(\prod_{m=1}^{M}\prod_{n=1}^{N_m}\alpha_{m,n}\sum_{i=1,i\neq m}^{M}P_i 2^{-\frac{B_{m,n}}{N_t-1}}\right)^{\frac{1}{K}}$$

$$= K\left(2^{-\frac{\sum_{m=1}^{M}\sum_{n=1}^{N_m}B_{m,n}}{N_t-1}}\right)^{\frac{1}{K}}\left(\prod_{m=1}^{M}\prod_{n=1}^{N_m}\alpha_{m,n}\sum_{i=1,i\neq m}^{M}P_i\right)^{\frac{1}{K}}$$

$$= K\left(2^{-\frac{B_{tot}}{N_t-1}}\right)^{\frac{1}{K}}\left(\prod_{m=1}^{M}\prod_{n=1}^{N_t}\alpha_{m,n}\sum_{i=1,i\neq m}^{M}P_i\right)^{\frac{1}{K}}, \quad (3.22)$$

where the equality holds true only when $\alpha_{m,n}\sum_{i=1,i\neq m}^{M}P_i 2^{-\frac{B_{m,n}}{N-t-1}}$, $\forall m,n$ are equal.
In other words, the objective function in (3.21) can be minimized while satisfying
the following condition:

$$\alpha_{m,n}\sum_{i=1,i\neq m}^{M}P_i 2^{-\frac{B_{m,n}}{N_t-1}} = \left(2^{-\frac{B_{tot}}{N_t-1}}\right)^{\frac{1}{K}}\left(\prod_{m=1}^{M}\prod_{n=1}^{N_m}\alpha_{m,n}\sum_{i=1,i\neq m}^{M}P_i\right)^{\frac{1}{K}}. \quad (3.23)$$

Hence, based on the relaxed optimization problem, the optimal number of feedback
bits for the $\text{UE}_{m,n}$ is given by

$$B_{m,n} = \frac{B_{\text{tot}}}{K} - \frac{1}{K}\sum_{i=1}^{M}\sum_{j=1}^{N_i}\log_2\left(\alpha_{i,j}\sum_{l=1,l\neq i}^{M}P_l\right) + \log_2\left(\alpha_{m,n}\sum_{l=1,l\neq m}^{M}P_l\right).$$

$$(3.24)$$

Given channel statistical information and transmit power allocation, it is easy to
determine the feedback distribution according to (3.24). Note that there exists an
integer constraint on the number of feedback bits in practice, so we should utilize
the maximum integer that is not larger than $B_{m,n}$ in (3.24), i.e., $\lfloor B_{m,n}\rfloor$, $\forall m,n$.

Remarks The number of feedback bits distributed to the $\text{UE}_{m,n}$ is determined by
the average inter-cluster interference generated by the $\text{UE}_{m,n}$ with respect to the
average inter-cluster interference of each UE. In other words, if one UE generates
more inter-cluster interference, it would be allocated with more feedback bits, so as
to facilitate a more accurate ZFBF to minimize the total interference.

3.3.4 Mode Selection

As discussed above, the performance of the massive access system is limited by both inter-cluster and intra-cluster interference. Although ZFBF at the BS and SIC at the UEs are jointly applied, there still exists residual interference. Intuitively, the strength of the residual interference mainly relies on the number of clusters and the number of UEs in each cluster. For instance, increasing the number of UEs in each cluster might reduce the inter-cluster interference, but also results in an increase in intra-cluster interference. Thus, it is desired to dynamically adjust the transmission mode, including the number of clusters and the number of UEs in each cluster, according to channel conditions and system parameters. For dynamic mode selection, we have the following lemma:

Lemma 2 *If the BS has no CSI about the downlink, it is optimal to set $M = 1$. On the other hand, if the BS has perfect CSI about the downlink, $N_m = 1$ is the best choice.*

Proof First, if there is no CSI, namely $\rho_{m,n} = 0, \forall m, n$, ZFBF cannot be utilized to mitigate the inter-cluster interference. If all the UEs belong to one cluster, interference can be mitigated as much as possible by SIC. In the case of perfect CSI at the BS, ZFBF can completely the interference. Thus, it is optimal to arrange one UE in one cluster.   □

In above, we consider two extreme scenarios of no and perfect CSI at the BS, respectively. In practice, the BS has partial CSI based on channel quantization feedback. Thus, we propose to dynamically choose the transmission mode for maximizing the sum of average transmission rate. For ease of implementation, we assume that the clusters have the same number of UEs N. Thus, transmission mode selection is equivalent to an optimization problem below:

$$J_3 : \max_{M,N} \sum_{m=1}^{M} \sum_{n=1}^{N} R_{m,n} \tag{3.25}$$

$$\text{s.t.} \quad \text{C5}: \ MN = K,$$

$$\text{C6}: \ M > 0,$$

$$\text{C7}: \ N > 0.$$

J_3 is also an integer programming problem, so it is difficult to obtain the closed-form solution. Under this condition, it is feasible to get the optimal solution by numerical search and the search complexity is $O(M^N)$. In order to control the complexity of SIC, the number of UEs in one cluster is usually small, e.g., $N = 2$. Therefore, the complexity of numerical search is acceptable.

3.3.5 Joint Optimization Scheme

In fact, transmit power, feedback bits and transmission mode are coupled, and determine the performance together. Therefore, it is better to jointly optimize these variables, so as to further improve the performance of the massive access systems. For example, given a transmission mode, it is easy to first allocate transmit power according to (3.19), and then distribute feedback bits according to (3.24). Finally, we can select an optimal transmission mode with the largest sum rate. The complexity of the joint optimization is mainly determined by the mode selection. As mentioned above, if the number of UEs in one cluster is small, the complex of mode selection is acceptable.

3.4 Asymptotic Analysis

In order to provide insightful guidelines for system design, we now pursue an asymptotic analysis on the average sum rate of the system. In particular, two extreme cases are studied, namely, interference limited and noise limited.

3.4.1 Interference Limited Case

With loss of generality, we let $P_{m,n} = \theta_{m,n} P_{tot}, \forall m, n$, where $0 < \theta_{m,n} < 1$ is a power allocation factor. For instance, $\theta_{m,n}$ is equal to

$$\frac{\left(\sum\limits_{v=1,v\neq m}^{M} \sum\limits_{j=1}^{N_v} \alpha_{v,j}(1-\rho_{v,j}) \right)^{-1}}{N_m \left(\sum\limits_{l=1}^{M} \left(\sum\limits_{v=1,v\neq l}^{M} \sum\limits_{j=1}^{N_v} \alpha_{v,j}(1-\rho_{v,j}) \right)^{-1} \right)}$$

in the proposed power allocation scheme in

Section 3.3.2. If the total power P_{tot} is large enough, the noise term of SINR in (3.6) is negligible. In this case, with the help of [23, Eq. (4.3311)], the average transmission rate of the kth UE $(k > 1)$ in the mth cluster reduces to

$$R_{m,n} = \frac{1}{\ln(2)} \sum_{i=1}^{M} \varXi_M \left(i, \{\eta_{m,n}^q\}_{q=1}^{M} \right) \ln(\eta_{m,n}^i)$$

$$- \frac{1}{\ln(2)} \sum_{i=1}^{M} \varXi_M \left(i, \{\beta_{m,n}^v\}_{v=1}^{M} \right) \ln(\beta_{m,n}^i), \tag{3.26}$$

where we have also used the fact that

$$\sum_{i=1}^{M} \varXi_M\left(i, \{\eta_{m,n}^q\}_{q=1}^M\right) = \sum_{i=1}^{M} \varXi_M\left(i, \{\beta_{m,n}^v\}_{v=1}^M\right) = 1. \tag{3.27}$$

Similarly, the asymptotic average transmission rate of the 1st UE in the mth UE can be obtained as

$$R_{n,1} = \frac{1}{\ln(2)} \sum_{i=1}^{M} \varXi_M\left(i, \{\eta_{m,1}^q\}_{q=1}^M\right) \ln\left(\eta_{m,1}^i\right)$$

$$- \frac{1}{\ln(2)} \sum_{i=1}^{M-1} \varXi_{M-1}\left(i, \{\beta_{m,1}^v\}_{v=1}^{M-1}\right) \ln\left(\beta_{m,1}^i\right). \tag{3.28}$$

Combining (3.26) and (3.28), we have the following important result:

Theorem 1 *In the region of high transmit power, the average transmission rate is independent of P_{tot}, and there exists a performance ceiling regardless of P_{tot}, i.e., once P_{tot} is larger than a saturation point, the average transmission rate will not increase further even the transmit power increases.*

Proof According to the definitions, $\eta_{m,n}^i$ and $\beta_{m,n}^i$ can be rewritten as $\eta_{m,n}^i = \omega_{m,n}^i P_{tot}$ and $\beta_{m,n}^i = \psi_{m,n}^i P_{tot}$, where

$$\omega_{m,n}^i = \begin{cases} \alpha_{m,n} \sum_{j=1}^{n} \theta_{i,j} & \text{if } i = m \\ \alpha_{m,n}(1 - \rho_{m,n}) \sum_{l=1}^{N_i} \theta_{i,l} & \text{if } i \neq m \end{cases},$$

and

$$\psi_{m,n}^i = \begin{cases} \alpha_{m,n} \sum_{j=1}^{n-1} \theta_{i,j} & \text{if } i = m \\ \alpha_{m,n}(1 - \rho_{m,n}) \sum_{l=1}^{N_i} \theta_{i,l} & \text{if } i \neq m \end{cases},$$

respectively. Thus, $\varXi_M\left(i, \{\eta_{m,n}^q\}_{q=1}^M\right)$ and $\varXi_M\left(i, \{\beta_{m,n}^v\}_{v=1}^M\right)$ are independent of P_{tot}. Hence, $R_{m,n}$ in (3.26) can be transformed as

$$R_{m,n} = \frac{1}{\ln(2)} \sum_{i=1}^{M} \varXi_M\left(i, \{\eta_{m,n}^q\}_{q=1}^M\right) (\ln(P_{tot}) + \ln(\omega_{m,n}^i))$$

$$- \frac{1}{\ln(2)} \sum_{i=1}^{M} \varXi_M\left(i, \{\beta_{m,n}^v\}_{v=1}^M\right) (\ln(P_{tot}) + \ln(\psi_{m,n}^i))$$

$$= \frac{1}{\ln(2)} \sum_{i=1}^{M} \varXi_M \left(i, \{\eta_{m,n}^q\}_{q=1}^{M}\right) \ln(\omega_{m,n}^i)$$

$$- \frac{1}{\ln(2)} \sum_{i=1}^{M} \varXi_M \left(i, \{\beta_{m,n}^v\}_{v=1}^{M}\right) \ln(\psi_{m,n}^i), \tag{3.29}$$

where Eq. (3.29) follows the fact that $\sum_{i=1}^{M} \varXi_M \left(i, \{\eta_{m,n}^q\}_{q=1}^{M}\right) = \sum_{i=1}^{M} \varXi_M$ $\left(i, \{\beta_{m,n}^v\}_{v=1}^{M}\right) = 1$. Similarly, we can rewrite $R_{m,1}$ in (3.28) as

$$R_{m,1} = \frac{1}{\ln(2)} \sum_{i=1}^{M} \varXi_M \left(i, \{\eta_{m,1}^q\}_{q=1}^{M}\right) \ln\left(\omega_{m,1}^i\right)$$

$$- \frac{1}{\ln(2)} \sum_{i=1}^{M-1} \varXi_{M-1} \left(i, \{\beta_{m,1}^v\}_{v=1}^{M-1}\right) \ln\left(\psi_{m,1}^i\right), \tag{3.30}$$

where

$$\omega_{m,1}^i = \begin{cases} \alpha_{m,1}\theta_{i,1} & \text{if } i = m \\ \alpha_{m,1}(1-\rho_{m,1}) \sum_{l=1}^{N_i} \theta_{i,l} & \text{if } i \neq m \end{cases},$$

and

$$\psi_{m,1}^i = \begin{cases} \alpha_{m,1}(1-\rho_{m,1}) \sum_{l=1}^{N_i} \theta_{i,l} & \text{if } i < m \\ \alpha_{m,1}(1-\rho_{m,1}) \sum_{l=1}^{N_{i+1}} \theta_{i+1,l} & \text{if } i \geq m \end{cases}.$$

Note that both (3.29) and (3.30) are regardless of P_{tot}, which proves Theorem 1. □

Now, we investigate the relation between the performance ceiling in Theorem 1 and the CSI accuracy $\rho_{m,n}$. First, we consider $R_{m,n}$ with $n > 1$. As $\rho_{m,n}$ asymptotically approaches 1, the inter-cluster interference is negligible. Then, $R_{m,n}$ can be further reduced as

$$R_{m,n}^{\text{ideal}} = \text{E}\left[\log_2\left(\alpha_{m,n}|\mathbf{h}_{m,n}^H\mathbf{w}_m|^2 \sum_{j=1}^{n} P_{m,j}\right)\right] - \text{E}\left[\log_2\left(\alpha_{m,n}|\mathbf{h}_{m,n}^H\mathbf{w}_m|^2 \sum_{j=1}^{n-1} P_{m,j}\right)\right]$$

$$= \log_2\left(\frac{\sum_{j=1}^{n} \omega_{m,j}}{\sum_{j=1}^{n-1} \psi_{m,j}}\right). \tag{3.31}$$

It is found that even with perfect CSI, the average transmission rate for the $(n > 1)$th UE is still upper bounded. The bound $\log_2 \left(\dfrac{\sum\limits_{j=1}^{n} \omega_{m,j}}{\sum\limits_{j=1}^{n-1} \psi_{m,j}} \right)$ is completely determined by channel conditions, and thus cannot be increased via power allocation. Differently, for the 1st UE, if the CSI at the BS is sufficiently accurate, the SINR $\gamma_{m,1}$ becomes high. As a result, the constant term 1 in the rate expression is negligible, and thus the average transmission rate can be approximated as

$$
\begin{aligned}
R_{m,1} \approx \;& \mathrm{E}\left[\log_2 \left(\frac{\alpha_{m,1}|\mathbf{h}_{m,1}^{H}\mathbf{w}_m|^2 P_{m,1}}{\alpha_{m,1}(1-\rho_{m,1}) \sum\limits_{i=1,i\neq m}^{M} |\mathbf{e}_{m,1}^{H}\mathbf{w}_i|^2 \sum\limits_{l=1}^{N_i} P_{i,l}} \right) \right] \\
= \;& \underbrace{\mathrm{E}\left[\log_2 \left(\alpha_{m,1}|\mathbf{h}_{m,1}^{H}\mathbf{w}_m|^2 P_{m,1} \right) \right]}_{\text{Ideal average rate}} \\
& - \underbrace{\mathrm{E}\left[\log_2 \left(\alpha_{m,1}(1-\rho_{m,1}) \sum\limits_{i=1,i\neq m}^{M} |\mathbf{e}_{m,1}^{H}\mathbf{w}_i|^2 \sum\limits_{l=1}^{N_i} P_{i,l} \right) \right]}_{\text{Rate loss due to imperfect CSI}}.
\end{aligned} \tag{3.32}
$$

In (3.32), the first term is the ideal average transmission rate with perfect CSI, and the second one is rate loss caused by imperfect CSI. We first check the term of the ideal average transmission rate, which is given by

$$
\begin{aligned}
R_{m,1}^{\text{ideal}} &= \mathrm{E}\left[\log_2 \left(\alpha_{m,1} P_{tot}\theta_{m,1}|\mathbf{h}_{m,1}^{H}\mathbf{w}_m|^2 \right) \right] \\
&= \log_2 \left(\alpha_{m,1} P_{tot}\theta_{m,1} \right) - \frac{C}{\ln(2)}.
\end{aligned} \tag{3.33}
$$

Note that if there is perfect CSI at the BS, the average transmission rate of the 1st UE increases proportionally to $\log_2(P_{tot})$ without a bound. However, as seen in (3.31), the $(n > 1)$th UE has an upper bounded rate under the same condition, which reconfirms the claim in Lemma 2 that it is optimal to arrange one UE in each cluster in presence of perfect CSI. Then, we investigate the rate loss due to imperfect CSI, which can be expressed as

$$
\begin{aligned}
R_{n,1}^{\text{loss}} =\;& \mathrm{E}\left[\log_2 \left(\alpha_{m,1}(1-\rho_{m,1}) P_{tot} \sum\limits_{i=1,i\neq m}^{M} |\mathbf{e}_{m,1}^{H}\mathbf{w}_i|^2 \sum\limits_{t=1}^{N_i} \theta_{i,t} \right) \right] \\
=\;& \log_2 \left(\alpha_{m,1}(1-\rho_{m,1}) P_{tot} \right) \\
& - \frac{1}{\ln(2)} \sum\limits_{i=1}^{M-1} \mathcal{Z}_{M-1}\left(i, \{\mu_{m,1}^{v}\}_{v=1}^{M-1} \right) \left(C - \ln\left(\mu_{m,1}^{i} \right) \right),
\end{aligned} \tag{3.34}
$$

where

$$\mu_{m,1}^{v} = \begin{cases} \sum_{l=1}^{N_v} \theta_{v,l} & \text{if } v < m \\ \sum_{l=1}^{N_{v+1}} \theta_{v+1,l} & \text{if } v \geq m \end{cases}.$$

Given a $\rho_{m,1}$, the rate loss $R_{m,1}^{loss}$ enlarges as the total transmit power P_{tot} increases. In order to keep the same rate of increase to the ideal rate $R_{m,1}^{ideal}$, the CSI accuracy $\rho_{m,1}$ should satisfy the following theorem:

Theorem 2 *Only when $(1 - \rho_{m,1})P_{tot}$ is equal to a constant ε, the average transmission rate of the 1st UE in the mth cluster with imperfect CSI remains a fixed gap with respect to the ideal rate. Specifically, the number of feedback bits should satisfy the condition that $B_{m,1} = (N_t - 1)\log_2(P_{tot}/\varepsilon)$.*

Proof The proof is intuitively. By substituting $\rho_{m,1} = 1 - 2^{-\frac{B_{m,1}}{N_t-1}}$ into $(1 - \rho_{m,1})P_{tot} = \varepsilon$, we can get $B_{m,1} = (N_t - 1)\log_2(P_{tot}/\varepsilon)$, which proves Theorem 2. □

Remarks For the CSI accuracy at the BS, $\frac{B_{m,1}}{N_t-1}$, namely spatial resolution, is a crucial factor. Specifically, given a requirement on CSI accuracy, it is possible to reduce the feedback bits by decreasing the number of antennas N_t. Yet, in order to fulfill the spatial degrees of freedom for ZFBF at the BS, N_t must be not smaller than $(N - 1)K + 1$. This is because the beam $\mathbf{w}_m$ for the mth cluster should be in the null space of the channels for the UEs in the other $M - 1$ clusters.

Furthermore, substituting (3.33) and (3.34) into (3.32), we have

$$R_{m,1} \approx -\log_2(1 - \rho_{m,1}) + \log_2(\theta_{m,1})$$
$$-\sum_{i=1}^{M-1} \Xi_{M-1}\left(i, \{\mu_{m,1}^{v}\}_{v=1}^{M-1}\right)\log_2\left(\mu_{m,1}^{i}\right). \tag{3.35}$$

Given a power allocation scheme, it is interesting that the bound of $R_{m,1}$ is independent of channel conditions. As analyzed above, it is possible to improve the average rate by improving the CSI accuracy. In particular, we have the following lemma:

Lemma 3 *At the high power region with a large number of feedback bits, the average rate of the 1st UE increases linearly as the number of feedback bits increases.*

Proof Replacing $\rho_{m,1}$ in (3.35) with $\rho_{m,1} = 1 - 2^{-\frac{B_{m,1}}{N_t-1}}$, $R_{m,1}$ is transformed as

$$R_{m,1} \approx \frac{B_{m,1}}{N_t - 1} + \log_2(\theta_{m,1}) - \sum_{i=1}^{M-1} \varXi_{M-1}\left(i, \{\mu_{m,1}^v\}_{v=1}^{M-1}\right) \log_2\left(\mu_{m,1}^i\right),$$

$$(3.36)$$

which yields Lemma 3.   □

3.4.2 Noise Limited Case

If the interference term is negligible with respect to the noise term due to a low transmit power, then the SINR $\gamma_{m,n}$, $\forall m, n$ is reduced as

$$\gamma_{m,n} = \alpha_{m,n}|\mathbf{h}_{m,n}^H \mathbf{w}_m|^2 P_{m,n}, \tag{3.37}$$

which is equivalent to the interference-free case. As discussed earlier, $|\mathbf{h}_{m,n}^H \mathbf{w}_m|^2$ is $\chi^2(2)$ distributed, then the average transmission rate can be computed as

$$R_{m,n} = \int_0^\infty \log_2\left(1 + P_{m,n}\alpha_{m,n}x\right) \exp(-x)dx$$

$$= -\exp\left(\frac{1}{P_{m,n}\alpha_{m,n}}\right) E_i\left(-\frac{1}{P_{m,n}\alpha_{m,n}}\right). \tag{3.38}$$

Note that Eq. (3.38) is independent of the CSI accuracy, thus it is unnecessary to carry out CSI feedback in this scenario. Since both intra-cluster interference and inter-cluster interference are negligible, ZFBF at the BS and SIC at the UEs are not required, and all optimization schemes asymptotically approach the same performance.

3.5 Numerical Results

To evaluate the performance of the proposed massive access technique, we present several simulation results under different scenarios. For convenience, we set $N_t = 6$, $M = 3$, $N = 2$, $B_{tot} = 12$, while $\alpha_{m,n}$ and $\rho_{m,n}$ are given in Table 3.1 for all simulation scenarios without extra specification. In addition, we use SNR (in dB) to represent $10 \log_{10} P_{tot}$.

First, we verify the accuracy of the derived theoretical expressions. As seen in Fig. 3.2, the theoretical expressions for both the 1st and the 2nd UEs in the 1st

Table 3.1 Parameter table for $(\alpha_{m,n}, \rho_{m,n})$, $\forall m \in [1, 3]$ and $n \in [1, 2]$

m \ n	1	2
1	(1.00, 0.90)	(0.10, 0.70)
2	(0.95, 0.85)	(0.20, 0.75)
3	(0.90, 0.80)	(0.15, 0.80)

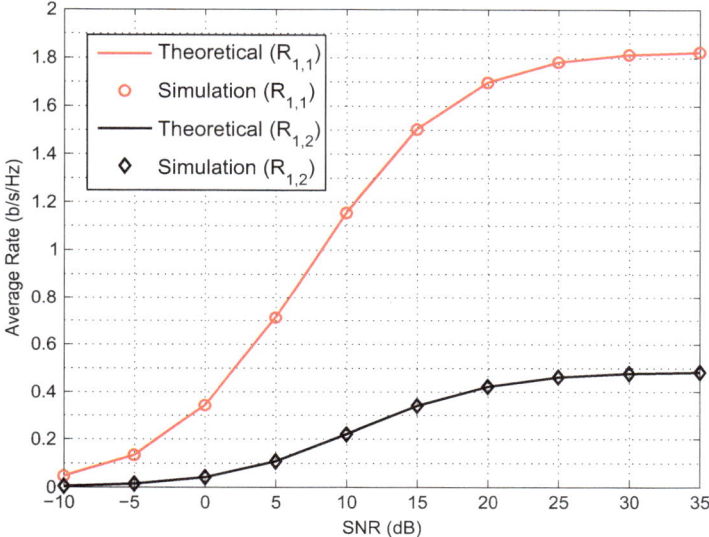

Fig. 3.2 Comparison of theoretical expressions and simulation results

cluster well coincide with the simulation results in the whole SNR region, which confirms the high accuracy. As the principle of NOMA implies, the 1st UE performs better than the second UE. At high SNR, the average rates of the both UEs are asymptotically saturated, which proves Theorem 1 again.

Secondly, we compare the proposed power allocation scheme with the equal power allocation scheme and the fixed power allocation scheme proposed in [25]. Note that the fixed power allocation scheme distributes the power with a fixed ratio 1:4 between the two UEs in a cluster so as to facilitate the SIC. It is found in Fig. 3.3 that the proposed power allocation scheme offers an obvious performance gain over the two baseline schemes, especially in the medium SNR region. Note that practical communication systems in general operate at medium SNR, thus the proposed scheme is able to achieve a given performance requirement with a lower SNR. As the SNR increases, the proposed scheme and the equal allocation scheme achieve the same saturated sum rate, but the fixed allocation scheme has a clear performance loss.

Next, we examine the advantage of feedback allocation for the massive system with equal power allocation, cf. Fig. 3.4. As analyzed above, at very low SNR, namely the noise-limited case, the average rate is independent of CSI accuracy,

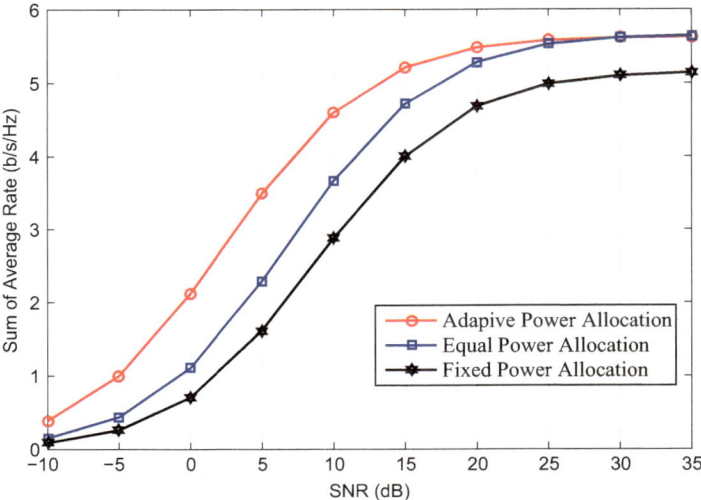

Fig. 3.3 Performance comparison of different power allocation schemes

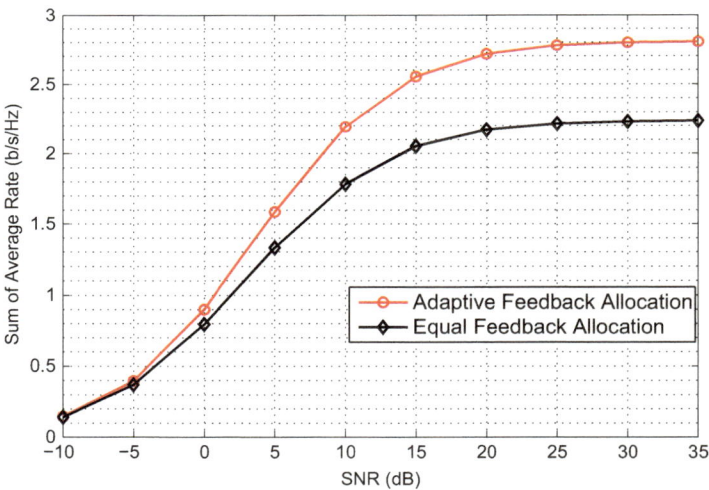

Fig. 3.4 Performance comparison of different feedback allocation schemes

and thus the two schemes asymptotically approach the same sum rate. As SNR increases, the proposed feedback allocation scheme achieves a larger performance gain. Similarly, at high SNR, both the two schemes are saturated, and the proposed scheme obtains the largest performance gain. For instance, at SNR= 30 dB, there is a gain of more than 0.5 b/s/Hz. Furthermore, we investigate the impact of the total number of feedback bits on the average rates of different UEs at SNR= 35 dB. As

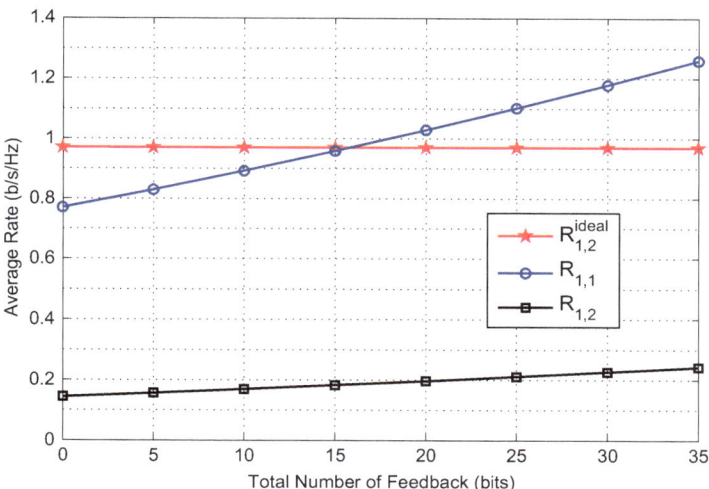

Fig. 3.5 Asymptotic performance with a large number of feedback bits

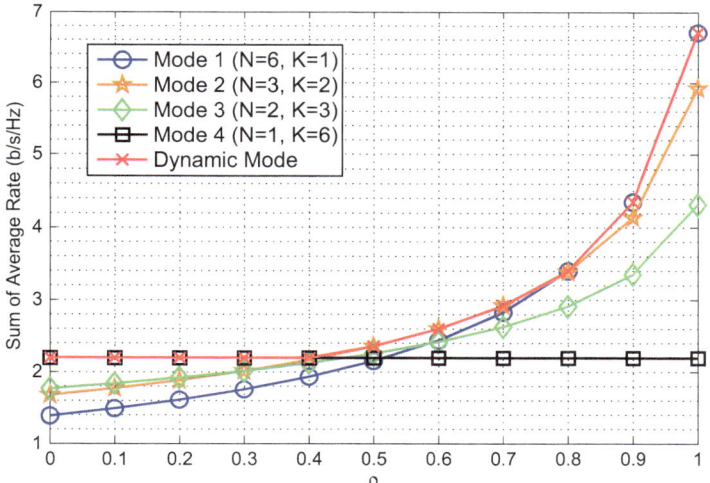

Fig. 3.6 Performance comparison of different transmission modes

shown in Fig. 3.5, the performance of the 1st UE is clearly better than that of the 2nd UE. Moreover, the average rate of the 1st UE is nearly a linear function of the number of feedback bits, which reconfirms the claims of Lemma 3.

Then, we investigate the impact of transmission mode on the performance of the massive access systems at SNR= 10 dB with equal power allocation in Fig. 3.6. To concentrate on the impact of transmission mode, we set the same CSI accuracy of all downlink channels as ρ. Note that we consider four fixed transmission modes under the same channel conditions in the case of 6 UEs in total. Consistent with the claims in Lemma 2, mode 4 with $M = 1$ and $N = 6$ achieves the largest sum rate at low CSI accuracy, while mode 1 with $N = 6$ and $K = 1$ performs best at high CSI accuracy. In addition, it is found that at medium CSI accuracy, mode 2 with $M = 3$ and $N = 2$ is optimal, since it is capable to achieve a best balance between intra-cluster interference and inter-cluster interference. Thus, we propose to dynamically select the transmission mode according to channel conditions and system parameters. As shown by the red line in Fig. 3.6, dynamic mode selection can always obtain the maximum sum rate.

Finally, we exhibit the superiority of the proposed joint optimization scheme for the massive systems at SNR= 10 dB. In addition, we take a fixed scheme based on NOMA and a time division multiple access (TDMA) based on OMA as baseline schemes. Specifically, the joint optimization scheme first distributes the transmit power with equal feedback allocation, then allocates the feedback bits based on the distributed power, finally selects the optimal transmission mode. The fixed scheme always adopts the mode 2 ($M = 3, N = 2$) with equal power and feedback allocation. The TDMA equally allocates each time slot to the 6 UEs, and utilizes maximum ratio transmission (MRT) based on the available CSI at the BS to maximize the rate. For clarity of notation, we use ρ to denote the CSI accuracy based on equal feedback allocation. In other words, the total number of feedback bits is equal to $B_{tot} = -K * (N_t - 1) * \log_2(1 - \rho)$. As seen in Fig. 3.7, the fixed scheme performs better than the TDMA scheme at low and high CSI accuracy, and slightly worse at medium regime. However, the proposed joint optimization scheme performs much better than the two baseline schemes. Especially at high CSI accuracy, the performance gap becomes substantially large. For instance, there

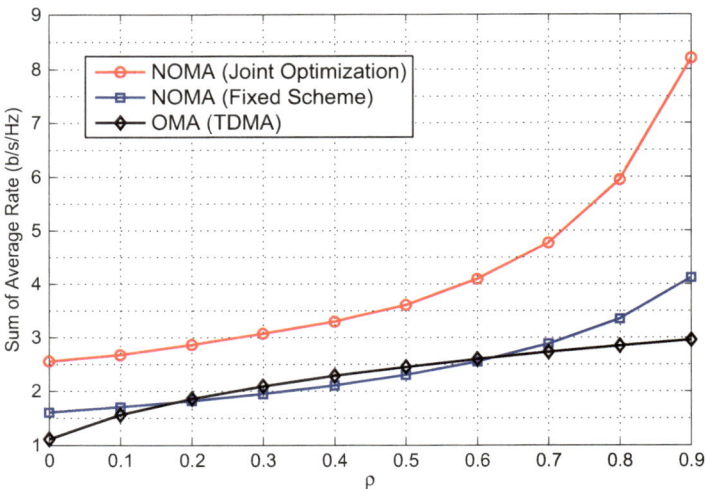

Fig. 3.7 Performance comparison of a joint optimization scheme and a fixed allocation scheme

is a performance gain of about 3 b/s/Hz at $\rho = 0.8$, and up to more than 5 b/s/Hz at $\rho = 0.9$. As analyzed in Lemma 2 and confirmed by Fig. 3.6, when ρ is larger than 0.8, which is a common CSI accuracy in practical systems, mode 2 is optimal for maximizing the system performance. Thus, the joint optimization scheme is reduced to joint power and feedback allocation, which requires only a very low complexity. Thus, the proposed NOMA scheme with joint optimization can achieve a good performance with low complexity, and it is a promising technique for future wireless communication systems.

3.6 Conclusion

In this chapter, we have provided a comprehensive solution for designing, analyzing, and optimizing a massive access system based on channel quantization feedback. First, we proposed a new framework for the massive access system. Then, we analyzed the performance, and derived exact closed-form expressions for average transmission rates. Afterwards, we optimized the three key parameters of massive access systems, i.e., transmit power, feedback bits, and transmission mode. Finally, we conducted asymptotic performance analysis, and obtained insights on system performance and design guidelines.

References

1. M. Shirvanimoghaddam, M. Dohler, S.J. Johnson, Massive non-orthogonal multiple access for cellular IoT: potentials and limitations. IEEE Commun. Mag. **55**(9), 55–61 (2017)
2. M. Shirvanimogaddam, M. Condoluci, M. Dohler, S.J. Johnson, On the fundamental limits of random non-orthogonal multiple access in cellular massive IoT. IEEE J. Sel. Areas Commun. **35**(10), 2238–2252 (2017)
3. D. Zhai, R. Zhang, L. Cai, B. Li, Y. Jiang, Energy-efficient user scheduling and power allocation for NOMA based wireless networks with massive IoT devices. IEEE Internet Things J. **5**(3), 1857–1868 (2018)
4. H. Weingarten, Y. Steinberg, S.S. Shamai, The capacity region of the Gaussian multiple-input multiple-output broadcast channel. IEEE Trans. Inf. Theory **52**(9), 3936–3964 (2006)
5. A.D. Dabbagh, D.J. Love, Precoding for multiple antenna Gaussian broadcast channels. IEEE Trans. Signal Process. **55**(7), 3837–3850 (2007)
6. Q. Sun, S. Han, I. Chin-Lin, Z. Pan, On the ergodic capacity of MIMO NOMA systems. IEEE Wirel. Commun. Lett. **4**(4), 405–408 (2015)
7. J. Choi, On the power allocation for MIMO-NOMA systems with layered transmission. IEEE Trans. Wirel. Commun. **15**(5), 3226–3237 (2016)
8. Z. Chen, Z. Ding, X. Dai, Beamforming for combating inter-cluster and intra-cluster interference in hybrid NOMA systems. IEEE Access **4**, 4452–4463 (2016)
9. Z. Ding, R. Schober, H.V. Poor, A general MIMO framework for NOMA downlink and uplink transmission based on signal alignment. IEEE Trans. Wirel. Commun. **15**(6), 4438–4454 (2016)
10. K. Higuchi, Y. Kishyama, Non-orthogonal access with random beamforming and intra-beam SIC for cellular MIMO downlink, in *Proceedings of the IEEE Vehicular Technology Conference (VTC-Fall)*, Las Vegas, Sept 2013, pp. 1–5

11. Z. Ding, F. Adachi, H.V. Poor, The application of MIMO to non-orthogonal multiple access. IEEE Trans. Wirel. Commun. **15**(1), 537–552 (2016)
12. X. Chen, Z. Zhang, C. Zhong, R. Jia, On the design of massive access, in *Proceedings of IEEE Wireless Communications & Signal Process (WCSP)*, Nanjing, Oct 2017, pp. 1–6
13. D.J. Love, R.W. Heath Jr., V.K.N. Lau, D. Gesbert, B.D. Rao, M. Andrews, An overview of limited feedback in wireless communication systems. IEEE J. Sel. Areas Commun. **26**(8), 1341–1365 (2008)
14. X. Chen, Z. Zhang, Exploiting channel angular domain information for precoder design in distributed antenna system. IEEE Trans. Signal Process. **58**(11), 5791–5801 (2010)
15. X. Chen, C. Yuen, Efficient resource allocation in rateless coded MU-MIMO cognitive radio network with QoS provisioning and limited feedback. IEEE Trans. Veh. Technol. **62**(1), 395–399 (2013)
16. N. Zhang, J. Wang, G. Kang, Y. Liu, Uplink non-orthogonal multiple access in 5G systems. IEEE Commun. Lett. **20**(3), 458–461 (2016)
17. Y. Liu, Z. Ding, M. Elkashlan, H.V. Poor, Cooperative non-orthogonal multiple access with simultaneous wireless information and power transfer. IEEE J. Sel. Areas Commun. **34**(4), 938–953 (2016)
18. N. Jindal, MIMO broadcast channels with finite-rate feedback. IEEE Trans. Inf. Theory **52**(11), 5045–5060 (2006)
19. X. Chen, L. Lei, Energy-efficient optimization for physical layer security in multi-antenna downlink networks with QoS guarantee. IEEE Commun. Lett. **17**(4), 637–640 (2013)
20. K. Saito, A. Benjebbour, Y. Kishiyama, Y. Okumura, T. Nakamura, Performance and design of SIC receiver for downlink NOMA with open-loop SU-MIMO, in *Proceedings of IEEE International Conference on Communication Workshop (ICCW)*, London, June 2015, pp. 1161–1165
21. K.K. Mukkavilli, A. Sabharwal, E. Erkip, B. Aazhang, On beamforming with finite rate feedback in multiple-antenna systems. IEEE Trans. Inf. Theory **49**(10), 2562–2579 (2003)
22. G.K. Karagiannidis, N.C. Sagias, T.A. Tsiftsis, Closed-form statistics for the sum of squared Nakagami-m variates and its application. IEEE Trans. Commun. **54**(8), 1353–1359 (2006)
23. I.S. Gradshteyn, I.M. Ryzhik, *Tables of Intergrals, Series, and Products* (Acedemic, USA, 2007)
24. M.B. Shahab, M.F. Kader, S.Y. Shin, On the power allocation of non-orthogonal multiple access for 5G wireless networks, in *Proceedings of International Conference Open Source Systems and Technologies (ICOSST)*, Lahore, Dec 2016, pp. 89–94
25. Z. Ding, Z. Yang, P. Fan, H.V. Poor, On the performance of non-orthogonal multiple access in 5G systems with randomly deployed users. IEEE Signal Process. Lett. **21**(12), 1501–1505 (2014)

Chapter 4
Massive Access with Channel Reciprocity

Abstract In this chapter, we propose a comprehensive fully non-orthogonal communication framework for cellular IoT in TDD mode. Firstly, we design a fully non-orthogonal communication scheme which consists of non-orthogonal channel estimation and non-orthogonal multiple access. Then, we analyze the performance of the proposed fully non-orthogonal communication, and derive a tight lower bound on the spectral efficiency in terms of key system parameters and channel conditions. Meanwhile, several novel insights are provided on spectral efficiency via asymptotic analysis in three important cases, i.e., a large number of base station (BS) antennas, a high BS transmit power, and perfect channel state information (CSI) at the BS. Finally, we optimize the performance of the proposed fully non-orthogonal communication and present two simple but efficient optimization algorithms for maximizing the weighted sum of spectral efficiency. Extensive simulation results validate the effectiveness of the proposed schemes.

4.1 Introduction

Massive MIMO has been widely regarded as a promising technique to support massive access in the 5G cellular IoT [1, 2]. To unlock the potential of massive MIMO to enable spectrally-efficient massive access, partial CSI should be available at the BS [3, 4]. However, the acquisition of CSI at the BS in massive MIMO systems is a challenging task, especially for the scenario of massive access. First, in frequency division duplex (FDD) systems, the required signalling overhead for CSI feedback is prohibitive. This is because the amount of CSI feedback should be proportional to the number of BS antennas and the number of users [5, 6]. As a result, massive MIMO systems are often suggested to operate in time division duplex (TDD) mode [7, 8]. By exploiting channel reciprocity of TDD systems, the BS is able to obtain the CSI of the downlink channels through estimating the uplink channels based on orthogonal training sequences. To maintain pairwise orthogonality of training sequences, the length of training sequence must be longer than the number of users [9]. Then, in the context of massive access, the training

© The Author(s), under exclusive license to Springer Nature Singapore Pte Ltd. 2019   65
X. Chen, *Massive Access for Cellular Internet of Things Theory and Technique*,
SpringerBriefs in Electrical and Computer Engineering,
https://doi.org/10.1007/978-981-13-6597-3_4

sequence is usually very long and might exceed the length of the channel coherence time. As a consequence, the estimated CSI might be outdated. In other words, conventional orthogonal channel estimation methods, e.g., [10] and [11], are not applicable to the massive access systems. To this end, the idea of pilot reuse was proposed to reduce the length of training sequence in [12] at the expense of sacrificing the CSI accuracy.

Other than channel estimation, there is also a vital and practical issue in the procedure of multiple access for non-orthogonal communication systems. As discussed above, to facilitate NOMA, a key step at the receiver side is the SIC for mitigating the intra-cluster interference. A common assumption in previous related works is that the receiver is capable of perfectly decoding the weak interfering signals and completely cancelling them in the received signal before demodulating its desired signal [13–15]. However, in practical systems, SIC is not a trivial task. Especially for some simple wireless devices with limited computational capability, decoding errors might be inevitable during the SIC, resulting in imperfect interference cancellation. It is intuitive that imperfect SIC would lead to a performance degradation. For alleviating the impact of imperfect SIC, it is desired to model imperfect SIC, and then optimize the corresponding performance. However, as pointed out in [16], there is no in-depth research that provides a mathematical model of the effect of imperfect SIC on NOMA schemes.

To jointly handle both challenging issues in channel estimation and multiple access for massive connections, we provide a comprehensive solution, namely fully non-orthogonal communication. Generally speaking, the proposed communication is non-orthogonal not only during multiple access, but also in channel estimation. The contributions of this chapter are as follows:

1. This chapter proposes a fully non-orthogonal communication framework to solve the practical issues in massive access. First, non-orthogonal channel estimation is performed to shorten the length of training sequences. Second, non-orthogonal multiple access is optimized to alleviate the impact of imperfect SIC.
2. This chapter analyzes the performance of the proposed fully non-orthogonal communication and derives a closed-form expression for a lower bound on the spectral efficiency in terms of system parameters and channel conditions. Furthermore, the impacts of key system parameters on spectral efficiency are revealed via asymptotic analysis in three important cases, i.e., a large number of BS antennas, a high BS transmit power, and perfect CSI at the BS.
3. This chapter optimizes the fully non-orthogonal communication. According to the characteristics of fully non-orthogonal communication, this chapter provides two algorithms to optimize the power allocation for channel estimation and multiple access for maximizing the weighted sum of spectral efficiency, respectively.

The rest of this chapter is organized as follows: Sect. 4.2 gives a brief introduction of a cellular IoT and presents a general fully non-orthogonal communication framework. Section 4.3 analyzes the performance of the proposed fully non-orthogonal communication and obtains some insights on the spectral efficiency. Next, Sect. 4.4 provides some effective schemes for optimizing the performance of channel estimation and multiple access. Then, extensive simulation results are presented in Sect. 4.5 to validate the effectiveness of the proposed theories and schemes. Finally, Sect. 4.6 concludes the chapter.

Notations We use bold upper (lower) letters to denote matrices (column vectors), $(\cdot)^H$ to denote conjugate transpose, $E[\cdot]$ to denote expectation, $var(\cdot)$ to denote the variance, $\|\cdot\|$ to denote the L_2-norm of a vector, $\otimes$ to denote the Kronecker product, $vec(\cdot)$ to denote the vectorization of a matrix, and $|\cdot|$ to denote the absolute value.

4.2 System Model and Problem Formulation

Consider a single-cell cellular IoT operated in TDD mode,[1] where a base station (BS) with N_t antennas serves K single-antenna UEs simultaneously,[2] as shown in Fig. 4.1. Note that the number of BS antennas N_t is suggested to be quite large in future 5G mobile communication systems compared to existing systems, e.g., $N_t \geq 64$ [17]. In other words, the BS is equipped with a large-scale antenna array. To exploit the benefits of multiple-antenna for supporting massive access and improving spectral efficiency, the NOMA technique is adopted by combining user clustering and spatial beamforming. The users in the same cluster share a spatial beam for eliminating the inter-cluster interference, and SIC is carried out in a cluster to mitigate the intra-cluster interference. Without loss of generality, we also assume that the K UEs are partitioned into M clusters, and the mth cluster contains N_m UEs. In what follows, we introduce the key techniques of fully non-orthogonal communication in detail.

4.2.1 Non-orthogonal Channel Estimation

The CSI availability at the BS is a prerequisite of efficient massive access schemes. To facilitate the following presentation, we use $\alpha_{m,n}^{1/2}\mathbf{h}_{m,n}$ to represent the N_t-dimensional channel vector from the BS to the nth UE in the mth cluster (namely $\mathrm{UE}_{m,n}$), where $\alpha_{m,n}$ is the distance-dependent path loss and $\mathbf{h}_{m,n}$ is the channel small-scale fading vector. Following the seminal works [7, 10], we assume that each element of $\mathbf{h}_{m,n}$ is an independent and identically distributed (i.i.d.) zero mean and unit variance complex Gaussian distributed random variable due to rich

---

[1]The proposed fully non-orthogonal communication framework can be extended to a multi-cell scenario directly. We will study the multi-cell case later.

[2]In practical systems, due to the physical size limitation, the IoT device is in general equipped with a single antenna [7, 12]. If the UE has multiple antennas, some signal receiving schemes, e.g., antenna selection, can be adopted to enhance the performance. Then, it is equivalent to the case of a single-antenna UE.

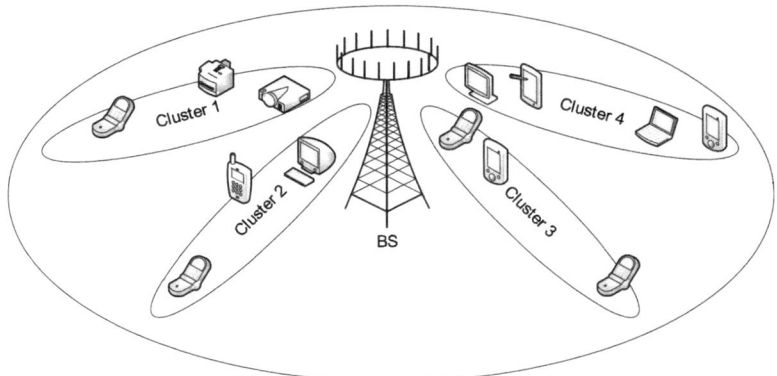

Fig. 4.1 A massive access model for the cellular IoT in TDD mode

scattering.[3] It is further assumed that $\alpha_{m,n}$ remains constant for a long time but $\mathbf{h}_{m,n}$ independently fades over time slots. Similar to existing works on TDD massive MIMO systems [7–9], the BS in the considered system directly obtains the CSI of the downlink channels by estimating the uplink channels by exploiting channel reciprocity. However, the conventional orthogonal channel estimation method requires very long pilot sequences in the scenario of massive access, resulting in outdated CSI. To overcome this challenge, we design a non-orthogonal channel estimation method for massive access.

At the beginning of each time slot, the UEs in the mth cluster send a common pilot sequence $\Phi_m \in \mathbf{C}^{1 \times \tau}$ with τ symbols over the uplink channels. Although the pilots in a cluster are the same, the pilots across the clusters are pairwise orthogonal, namely $\Phi_i \Phi_j^H = 0$ and $\Phi_i \Phi_i^H = 1, \forall i \neq j,$. To guarantee pairwise orthogonality, the length of pilot sequence τ must not be shorter than the number of clusters M, which is usually satisfied in practical systems. Compared to conventional orthogonal channel estimation methods, e.g., [10] and [11], the proposed non-orthogonal channel estimation can shorten N times of the length of training sequence. Therefore, even for the case of massive connections, the length of pilot sequence can be guaranteed to be less than the duration of channel coherence time. Based on the proposed non-orthogonal transmission, the received signal across τ symbols at the BS can be expressed as

---

[3]In practice, the large-scale antenna array at the BS might be correlated in some scenarios. In general, the channel from the BS to the $\text{UE}_{m,n}$ can be represented as $\mathbf{R}_{m,n}\mathbf{h}_{m,n}$, where $\mathbf{R}_{m,n}$ is the correlation matrix and $\mathbf{h}_{m,n}$ is the channel small-scale fading vector with i.i.d. zero mean and unit variance complex Gaussian distributed random variable. Since the correlation matrix $\mathbf{R}_{m,n}$ usually varies slowly compared to $\mathbf{h}_{m,n}$, it is reasonably assumed that $\mathbf{R}_{m,n}$ is known at both the BS and the UE. Mathematically, the impact of the correlation matrix $\mathbf{R}_{m,n}$ is equivalent to that of the path loss $\alpha_{m,n}$ in the case of i.i.d. channel, thus the proposed fully non-orthogonal transmission scheme is also applicable in the scenario of channel correlation.

$$\mathbf{Y} = \sum_{i=1}^{M} \sum_{j=1}^{N_m} \sqrt{\alpha_{i,j} Q_{i,j} \tau} \mathbf{h}_{i,j} \Phi_i + \mathbf{N}, \tag{4.1}$$

where $Q_{i,j}$ is the transmit power of pilot sequence from the $\mathrm{UE}_{i,j}$ and $\mathbf{N}$ is an AWGN matrix with unit variance entries. Since the UEs in a cluster utilize the same pilot sequence and share a transmit beam, we only need to estimate the effective CSI $\mathbf{h}_i$ for the ith cluster, where $i = 1, \cdots, M$. Let us take the estimation of $\mathbf{h}_m$ as an example. First, right-multiplexing $\mathbf{Y}$ by Φ_m^H, we have

$$\mathbf{Y}\Phi_m^H = \sum_{j=1}^{N_m} \sqrt{\alpha_{m,j} Q_{m,j} \tau} \mathbf{h}_{m,j} + \mathbf{N}\Phi_m^H$$

$$= \sqrt{\sum_{j=1}^{N_m} \alpha_{m,j} Q_{m,j} \tau} \mathbf{h}_m + \mathbf{N}\Phi_m^H, \tag{4.2}$$

where $\mathbf{h}_m = \dfrac{\sum_{n=1}^{N_m} \sqrt{\alpha_{m,n} Q_{m,n} \tau} \mathbf{h}_{m,n}}{\sqrt{\sum_{j=1}^{N_m} \alpha_{m,j} Q_{m,j} \tau}}$ is the effective CSI for the mth cluster. Then, by using minimum mean squared error (MMSE) channel estimation, the estimated CSI $\hat{\mathbf{h}}_m$ is given by

$$\hat{\mathbf{h}}_m = \frac{\sqrt{\sum_{n=1}^{N_m} \alpha_{m,n} Q_{m,n} \tau}}{1 + \sum_{j=1}^{N_m} \alpha_{m,j} Q_{m,j} \tau} \hat{\mathbf{y}}_m, \tag{4.3}$$

where

$$\hat{\mathbf{y}}_m = \sum_{j=1}^{N_m} \sqrt{\alpha_{m,j} Q_{m,j} \tau} \mathbf{h}_{m,j} + (\Phi_m \otimes \mathbf{I}_{N_t}) \mathbf{n}, \tag{4.4}$$

and $\mathbf{n} = \mathrm{vec}(\mathbf{N})$. Combing (4.3) and (4.4), for the $\mathrm{UE}_{m,n}$, the relation between the actual CSI $\mathbf{h}_{m,n}$ and the estimated CSI $\hat{\mathbf{h}}_m$ can be expressed as

$$\mathbf{h}_{m,n} = \sqrt{\rho_{m,n}} \hat{\mathbf{h}}_m + \sqrt{1 - \rho_{m,n}} \mathbf{e}_{m,n}, \tag{4.5}$$

where $\mathbf{e}_{m,n}$ is the channel estimation error vector with i.i.d. zero mean and unit variance complex Gaussian distributed entries, and is independent of $\hat{\mathbf{h}}_m$. The variable $\rho_{m,n} = \dfrac{\alpha_{m,n} Q_{m,n} \tau}{1 + \sum\limits_{j=1}^{N_m} \alpha_{m,j} Q_{m,j} \tau}$ is the correlation coefficient between $\mathbf{h}_{m,n}$ and $\hat{\mathbf{h}}_m$.

Note that a large $\rho_{m,n}$ means a high channel estimation accuracy.

Due to non-orthogonal channel estimation, the CSI accuracies among the users in a cluster are coupled. Without loss in generality, checking the correlation coefficients in the mth cluster, we have

$$\sum_{n=1}^{N_m} \rho_{m,n} = \sum_{n=1}^{N_m} \frac{\alpha_{m,n} Q_{m,n} \tau}{1 + \sum\limits_{j=1}^{N_m} \alpha_{m,j} Q_{m,j} \tau}$$

$$= 1 - \frac{1}{1 + \sum\limits_{j=1}^{N_m} \alpha_{m,j} Q_{m,j} \tau} \le 1, \qquad (4.6)$$

where $\sum\limits_{j=1}^{N_m} \alpha_{m,j} Q_{m,j} \tau$ is the total energy of the received pilots from the mth cluster.

It is found that the CSI accuracies among the users in a cluster are conflicted and the sum of the CSI accuracies is upper bounded by 1. Thus, for a given τ, one can improve the CSI accuracy by coordinating the transmit powers in a cluster.

Remark 1 It is possible to improve the CSI accuracy by assigning multiple orthogonal pilot sequences in a cluster. However, such a method requires a longer pilot sequence, resulting in a less duration for information transmission in each time slot. From the perspective of maximizing the overall performance of fully non-orthogonal communication systems, it is desired to select the optimal number of orthogonal pilot sequences in a cluster. Yet, an exhaustive search is required which leads to prohibitively large computational complexity. Therefore, in this chapter, we assume that orthogonal pilots used only across clusters as a first attempt to study the performance of fully non-orthogonal communication systems.

4.2.2 Non-orthogonal Multiple Access

Based on the CSI obtained by non-orthogonal channel estimation, we design a spectral-efficient massive non-orthogonal multiple access scheme. Specifically, superposition coding is carried out at the BS to combat the inter-cluster interference, while SIC is performed at the user to mitigate the intra-cluster interference, and thus improve the spectral efficiency. First, in a cluster, the BS constructs the transmit signal x_m as follow:

$$x_m = \sum_{n=1}^{N_m} \sqrt{P_{m,n}} s_{m,n}, \tag{4.7}$$

where $s_{m,n}$ and $P_{m,n}$ are the complex Gaussian distributed information signal of unit norm and transmit power for the $\text{UE}_{m,n}$, respectively. The transmit power $P_{n,n}$ can be optimized to coordinate the intra-cluster interference. Then, the BS constructs the total transmit signal $\mathbf{x}$ as below:

$$\mathbf{x} = \sum_{m=1}^{M} \mathbf{w}_m x_m. \tag{4.8}$$

where $\mathbf{w}_m$ is the transmit beam for the mth cluster. For a balance between system performance and implementation complexity, we adopt a match filtering (MF)-based beamforming design for the fully non-orthogonal communication with user clustering. Specifically, $\mathbf{w}_m$ is constructed as

$$\mathbf{w}_m = \frac{\hat{\mathbf{h}}_m}{\|\hat{\mathbf{h}}_m\|}. \tag{4.9}$$

Finally, the BS broadcasts the signal $\mathbf{x}$ over the downlink channel.

Remarks A multiple-antenna system adopting such a beamforming design can support an arbitrary number of beams, and each beam can be shared by an arbitrary number of users. Thus, the multiple-antenna system has a huge potential of supporting spectrally-efficient massive access. Actually, the transmit beam $\mathbf{w}_m$ can be designed from different viewpoints. For example, ZF beamforming can mitigate the inter-cluster interference and generalized eigenvector beamforming is able to maximize the signal-to-interference-plus-noise ratio (SINR) [18].

Without loss in generality, we consider the received signal $y_{m,n}$ at the $\text{UE}_{m,n}$, which is given by

$$
\begin{aligned}
y_{m,n} &= \alpha_{m,n}^{1/2} \mathbf{h}_{m,n}^H \mathbf{x} + n_{m,n} \\
&= \underbrace{\alpha_{m,n}^{1/2} \mathbf{h}_{m,n}^H \mathbf{w}_m \sqrt{P_{m,n}} s_{m,n}}_{\text{Desired signal}} + \underbrace{\sum_{i=1, i \neq n}^{N_m} \alpha_{m,n}^{1/2} \mathbf{h}_{m,n}^H \mathbf{w}_m \sqrt{P_{m,i}} s_{m,i}}_{\text{Intra-cluster interference}} \\
&\quad + \underbrace{\sum_{j=1, j \neq m}^{M} \sum_{i=1}^{N_j} \alpha_{m,n}^{1/2} \mathbf{h}_{m,n}^H \mathbf{w}_j \sqrt{P_{j,i}} s_{j,i}}_{\text{Inter-cluster interference}} + \underbrace{n_{m,n}}_{\text{AWGN}} ,
\end{aligned} \tag{4.10}
$$

where $n_{m,n}$ is additive white Gaussian noise (AWGN) with unit variance. Note that the second term on the right side of Eq. (4.10) is the intra-cluster interference caused by non-orthogonal transmission, which can be reduced by SIC. It is assumed that the effective channel gains in the mth cluster have the following order:

$$|\alpha_{m,1}^{1/2}\mathbf{h}_{m,1}^H\mathbf{w}_m|^2 \geq |\alpha_{m,2}^{1/2}\mathbf{h}_{m,2}^H\mathbf{w}_m|^2 \geq \cdots \geq |\alpha_{m,N}^{1/2}\mathbf{h}_{m,N}^H\mathbf{w}_m|^2. \qquad (4.11)$$

In practice, each UE knows its effective channel gain through channel estimation for coherent detection [19, 20]. Thus, the UEs can feed back the effective channel gains to the BS via the uplink and the BS determines the order of effective channel gains in each cluster which is informed to the UEs through the downlink. Based on the order of effective channel gains, the nth UE decodes the interfering signal and subtracts the corresponding interference from the Nth to $(n + 1)$th UE in sequence, and finally demodulates its desired signal $s_{m,n}$. If SIC is perfectly performed, the UE can completely cancel the intra-cluster interference from the UEs with weaker effective channel gains. However, in practical systems, due to the hardware limitation of the IoT device, the low signal quality, and other factors, the decoding error of the weak interference signal might occur. As a result, there exists residual interference from the weak UE after SIC. In this case, the post-SIC signal at the $\text{UE}_{m,n}$ is given by

$$y'_{m,n} = \alpha_{m,n}^{1/2}\mathbf{h}_{m,n}^H\mathbf{w}_m\sqrt{P_{m,n}}s_{m,n} + \sum_{i=1}^{n-1}\alpha_{m,n}^{1/2}\mathbf{h}_{m,n}^H\mathbf{w}_m\sqrt{P_{m,i}}s_{m,i}$$

$$+ \sum_{i=n+1}^{N_m}\alpha_{m,n}^{1/2}\mathbf{h}_{m,n}^H\mathbf{w}_m\sqrt{\eta_{m,n}P_{m,i}}s_{m,i} + \sum_{j=1,j\neq m}^{M}\sum_{i=1}^{N_j}\alpha_{m,n}^{1/2}\mathbf{h}_{m,n}^H\mathbf{w}_j\sqrt{P_{j,i}}s_{j,i} + n_{m,n},$$

$$(4.12)$$

where $\eta_{m,n} \leq 1$ denotes the coefficient of imperfect SIC at the $\text{UE}_{m,n}$, which can be obtained by long-term measurement.[4] If SIC is perfect, we have $\eta_{m,n} = 0$, and thus the intra-cluster interference from the weak users can be completely cancelled. In other words, perfect SIC in existing works is a special case of this chapter.

4.3 Analysis of Fully Non-orthogonal Communication

In this section, we first evaluate the performance of the proposed fully non-orthogonal communication in terms of spectral efficiency. Then, we analyze the asymptotical characteristics of spectral efficiency for revealing the impacts of key system parameters.

---

[4]By measuring a large number of samples in a long training time, the residual interference over i.i.d. Rayleigh channels can be accurately approximated using a Gaussian distribution due to the central limit theorem and its variance is a function of the received power [21]. Then, by comparing the powers of the residual interference and the received signal, the coefficient $\eta_{m,n}$ can be obtained.

4.3.1 Spectral Efficiency of Fully Non-orthogonal Communication

Without loss in generality, we focus on the analysis of the spectral efficiency for the nth user in the mth cluster, which is given by

$$R_{m,n} = \mathrm{E}[\log_2(1 + \gamma_{m,n})], \tag{4.13}$$

where $\gamma_{m,n}$ is the received SINR, which can be expressed as

$$\gamma_{m,n} = \frac{|\mathbf{h}_{m,n}^H \mathbf{w}_m|^2 \alpha_{m,n} P_{m,n}}{|\mathbf{h}_{m,n}^H \mathbf{w}_m|^2 \alpha_{m,n} \sum\limits_{i=1,i\neq n}^{N_m} \kappa_{m,i} P_{m,i} + \sum\limits_{j=1,j\neq m}^{M} \sum\limits_{i=1}^{N_j} |\mathbf{h}_{m,n}^H \mathbf{w}_j|^2 \alpha_{m,n} P_{j,i} + 1}, \tag{4.14}$$

and $\kappa_{m,i}$ is defined as

$$\kappa_{m,i} = \begin{cases} \eta_{m,n}, & \text{if } i > n, \\ 1, & \text{otherwise.} \end{cases} \tag{4.15}$$

Note that the computation of the expectation with respect to a complicated random variable $\gamma_{m,n}$ in (4.14) is not a trivial task. As a compromise, we focus on the derivation of a lower bound on the spectral efficiency. Specifically, we first rewrite the received signal $y'_{m,n}$ in (4.12) as follows:

$$y'_{m,n} = \mathrm{E}[\mathbf{h}_{m,n}^H \mathbf{w}_m \sqrt{\alpha_{m,n} P_{m,n}}] s_{m,n} + n'_{m,n}, \tag{4.16}$$

where $n'_{m,n}$ is an effective noise, which is given by

$$n'_{m,n} = I_{1,m,n} + I_{2,m,n} + I_{3,m,n} + n_{m,n}, \tag{4.17}$$

with

$$I_{1,m,n} = \left(\mathbf{h}_{m,n}^H \mathbf{w}_m \sqrt{\alpha_{m,n} P_{m,n}} - \mathrm{E}[\mathbf{h}_{m,n}^H \mathbf{w}_m \sqrt{\alpha_{m,n} P_{m,n}}]\right) s_{m,n}, \tag{4.18}$$

$$I_{2,m,n} = \mathbf{h}_{m,n}^H \mathbf{w}_m \sum\limits_{i=1,i\neq n}^{N_m} \sqrt{\alpha_{m,n} \kappa_{m,i} P_{m,i}} s_{m,i}, \tag{4.19}$$

and

$$I_{3,m,n} = \sum_{j=1,j\neq m}^{M} \sum_{i=1}^{N_j} \mathbf{h}_{m,n}^{H} \mathbf{w}_j \sqrt{\alpha_{m,n} P_{j,i}} s_{j,i}, \tag{4.20}$$

being the signal leakage, the intra-cluster interference, and the inter-cluster interference, respectively. Then, if it is assumed that $n'_{m,n}$ is an independent Gaussian noise, we can obtain a lower bound on the spectral efficiency as $\underline{R}_{m,n} = \log_2(1 + \underline{\gamma}_{m,n}) \leq R_{m,n}$ [10], where $\underline{\gamma}_{m,n}$ denotes an effective SINR, which is given by

$$\underline{\gamma}_{m,n} = \frac{\phi_{0,m,n}}{\phi_{1,m,n} + \phi_{2,m,n} + \phi_{3,m,n} + 1}, \tag{4.21}$$

with

$$\phi_{0,m,n} = \left| E\left[\mathbf{h}_{m,n}^{H} \mathbf{w}_m \sqrt{\alpha_{m,n} P_{m,n}} \right] \right|^2, \tag{4.22}$$

$$\phi_{1,m,n} = \mathrm{var}\left(\mathbf{h}_{m,n}^{H} \mathbf{w}_m \sqrt{\alpha_{m,n} P_{m,n}} \right), \tag{4.23}$$

$$\phi_{2,m,n} = \sum_{i=1,i\neq n}^{N_m} E\left[|\mathbf{h}_{m,n}^{H} \mathbf{w}_m \sqrt{\alpha_{m,n} \kappa_{m,i} P_{m,i}}|^2 \right], \tag{4.24}$$

and

$$\phi_{3,m,n} = \sum_{j=1,j\neq m}^{M} \sum_{i=1}^{N_j} E\left[\left| \mathbf{h}_{m,n}^{H} \mathbf{w}_j \sqrt{\alpha_{m,n} P_{j,i}} \right|^2 \right], \tag{4.25}$$

being the variances of the desired signal, the signal leakage, the residual intra-cluster interference, and the residual inter-cluster interference, respectively. Thus, the key of deriving the spectral efficiency is to compute the mean and variance in (4.22)–(4.25). In the following, we provide a detailed derivation based on the characteristics of fully non-orthogonal communication.

First, we analyze the term $E\left[\mathbf{h}_{m,n}^{H} \mathbf{w}_m \sqrt{\alpha_{m,n} P_{m,n}} \right]$ at the numerator of (4.21). Substituting $\mathbf{h}_{m,n} = \sqrt{\rho_{m,n}}\hat{\mathbf{h}}_m + \sqrt{1 - \rho_{m,n}}\mathbf{e}_{m,n}$ into the term, we have

$$E\left[\mathbf{h}_{m,n}^{H} \mathbf{w}_m \sqrt{\alpha_{m,n} P_{m,n}} \right] = \sqrt{\alpha_{m,n} P_{m,n}} E\left[\left(\sqrt{\rho_{m,i}}\hat{\mathbf{h}}_m^{H} + \sqrt{1 - \rho_{m,i}}\mathbf{e}_{m,n}^{H} \right) \mathbf{w}_m \right]$$

$$= \sqrt{\alpha_{m,n} P_{m,n} \rho_{m,n}} E\left[\hat{\mathbf{h}}_m^{H} \mathbf{w}_m \right] \tag{4.26}$$

$$= \sqrt{\alpha_{m,n} P_{m,n} \rho_{m,n}} E\left[\|\hat{\mathbf{h}}_m\| \right] \tag{4.27}$$

$$= \sqrt{\alpha_{m,n} P_{m,n} \rho_{m,n}} \frac{\Gamma(N_t + 1/2)}{\Gamma(N_t)}, \qquad (4.28)$$

where $\Gamma(x)$ denotes the Gamma function. Equation (4.26) holds true because $\mathbf{e}_{m,n}$ is independent of $\mathbf{w}_m$ and $\mathrm{E}\left[\mathbf{e}_{m,n}\right] = 0$, Eq. (4.27) replaces $\mathbf{w}_m$ with $\frac{\hat{\mathbf{h}}_m}{\|\hat{\mathbf{h}}_m\|}$ and Eq. (4.28) follows the fact that $\|\hat{\mathbf{h}}_m\|^2$ is χ^2 distributed with $2N_t$ degrees of freedom. In other words, $\mathrm{E}\left[\|\hat{\mathbf{h}}_m\|\right]$ is equal to $\frac{\Gamma(N_t+1/2)}{\Gamma(N_t)}$. Therefore, the power of the desired signal can be expressed as

$$\phi_{0,m,n} = \alpha_{m,n} P_{m,n} \rho_{m,n} \frac{\Gamma^2(N_t + 1/2)}{\Gamma^2(N_t)}. \qquad (4.29)$$

Then, we analyze the power of the signal leakage in the denominator as below:

$$\phi_{1,m,n} = \mathrm{var}\left(\mathbf{h}_{m,n}^H \mathbf{w}_m \sqrt{\alpha_{m,n} P_{m,n}}\right)$$

$$= \alpha_{m,n} P_{m,n} \mathrm{E}\left[\left|\mathbf{h}_{m,n}^H \mathbf{w}_m - \mathrm{E}\left[\mathbf{h}_{m,n}^H \mathbf{w}_m\right]\right|^2\right]$$

$$= \alpha_{m,n} P_{m,n} \left(\mathrm{E}\left[\left|\mathbf{h}_{m,n}^H \mathbf{w}_m\right|^2\right] - \left|\mathrm{E}\left[\mathbf{h}_{m,n}^H \mathbf{w}_m\right]\right|^2\right), \qquad (4.30)$$

where the first expectation term in (4.30) can be computed as

$$\mathrm{E}\left[\left|\mathbf{h}_{m,n}^H \mathbf{w}_m\right|^2\right] = \mathrm{E}\left[\left|\sqrt{\rho_{m,n}}\hat{\mathbf{h}}_m^H \mathbf{w}_m + \sqrt{1 - \rho_{m,n}}\mathbf{e}_{m,n}^H \mathbf{w}_m\right|^2\right]$$

$$= \rho_{m,n}\mathrm{E}\left[\left|\hat{\mathbf{h}}_m^H \mathbf{w}_m\right|^2\right] + (1 - \rho_{m,n})\mathrm{E}\left[\left|\mathbf{e}_{m,n}^H \mathbf{w}_m\right|^2\right], \qquad (4.31)$$

where Eq. (4.31) follows the fact that $\hat{\mathbf{h}}_m$ and $\mathbf{e}_{m,n}$ are independent of each other. As discussed above, $\left|\hat{\mathbf{h}}_m^H \mathbf{w}_m\right|^2 = \|\hat{\mathbf{h}}_m\|^2$ is χ^2 distributed with $2N_t$ degrees of freedom, thus we have $\mathrm{E}\left[\left|\hat{\mathbf{h}}_m^H \mathbf{w}_m\right|^2\right] = N_t$. Moreover, since $\mathbf{w}_m$ is independent of $\mathbf{e}_{m,n}$, $\left|\mathbf{e}_{m,n}^H \mathbf{w}_m\right|^2$ is χ^2 distributed with 2 degrees of freedom, and thus we have $\mathrm{E}\left[\left|\mathbf{e}_{m,n}^H \mathbf{w}_m\right|^2\right] = 1$. To sum up, it is able to obtain that

$$\mathrm{E}\left[\left|\mathbf{h}_{m,n}^H \mathbf{w}_m\right|^2\right] = \rho_{m,n} N_t + 1 - \rho_{m,n}. \qquad (4.32)$$

Substituting (4.28) and (4.32) into (4.30) yields

$$\phi_{1,m,n} = \alpha_{m,n} P_{m,n} \left(\rho_{m,n} \left(N_t - 1 - \frac{\Gamma^2(N_t + 1/2)}{\Gamma^2(N_t)} \right) + 1 \right). \tag{4.33}$$

Based on the above analysis, the powers of the intra-cluster interference and the inter-cluster interference in the denominator of (4.21) can be computed as

$$\phi_{2,m,n} = \sum_{i=1, i \neq n}^{N_m} \mathrm{E} \left[|\mathbf{h}_{m,n}^H \mathbf{w}_m \sqrt{\alpha_{m,n} \kappa_{m,i} P_{m,i}}|^2 \right]$$

$$= \alpha_{m,n} \left(\rho_{m,n} (N_t - 1) + 1 \right) \sum_{i=1, i \neq n}^{N_m} \kappa_{m,i} P_{m,i}, \tag{4.34}$$

and

$$\phi_{3,m,n} = \sum_{j=1, j \neq m}^{M} \sum_{i=1}^{N_j} \mathrm{E} \left[|\mathbf{h}_{m,n}^H \mathbf{w}_j \sqrt{\alpha_{m,n} P_{j,i}}|^2 \right]$$

$$= \sum_{j=1, j \neq m}^{M} \sum_{i=1}^{N_j} \alpha_{m,n} P_{j,i}, \tag{4.35}$$

respectively, where Eq. (4.35) holds true since $\left| \mathbf{h}_{m,n}^H \mathbf{w}_j \right|^2$ is χ^2 distributed with 2 degrees of freedom. Finally, substituting (4.28), (4.33), (4.34), and (4.35) into (4.21), we can obtain a closed-form expression for the lower bound on the spectral efficiency in a fully non-orthogonal communication, which is given by

$$\underline{R}_{m,n} = \log_2 \left(1 + \frac{\alpha_{m,n} P_{m,n} \rho_{m,n} \frac{\Gamma^2(N_t+1/2)}{\Gamma^2(N_t)}}{\left[\alpha_{m,n} (\rho_{m,n}(N_t - 1) + 1) \sum_{i=1}^{N_m} \kappa_{m,i} P_{m,i} \right. \\ \left. - \alpha_{m,n} P_{m,n} \rho_{m,n} \frac{\Gamma^2(N_t+1/2)}{\Gamma^2(N_t)} + \alpha_{m,n} \sum_{j=1, j \neq m}^{M} \sum_{i=1}^{N_j} P_{j,i} + 1 \right]} \right). \tag{4.36}$$

As will be verified by simulations, the lower bound on the spectral efficiency in (4.36) is tight at medium and high SNR regimes when the number of BS antennas is large. Thus, it is able to accurately evaluate the performance of the system even if there is an arbitrary number of users once the system parameters and channel conditions are known. Moreover, it is possible to adjust the parameters, i.e., CSI accuracy and transmit power, so as to meet a given performance requirement with a minimum resource consumption.

Remark 2 Each user's spectral efficiency is determined by its own CSI accuracy. However, according to the characteristics of the proposed non-orthogonal channel estimation in (4.6), the CSI accuracies among the users in a cluster are coupled. Hence, it is likely to optimize the CSI accuracies in a cluster, but not necessarily in the whole system.

Remark 3 Each user's spectral efficiency is affected by its belonging clusters' power allocation, but not the other clusters' power allocation. This is because the inter-cluster interference is directly determined by the total power of other clusters, namely $\sum_{j=1, j\neq m}^{M} \sum_{i=1}^{N_j} \alpha_{m,n} P_{j,i} = \alpha_{m,n} \sum_{j=1, j\neq m}^{M} P_j$, where P_j is the total transmit power for the jth cluster. Therefore, each cluster can independently perform power allocation for a given total transmit power of each cluster.

4.3.2   The Multi-cell Case

The considered single-cell fully non-orthogonal communication in the last sections can be easily extended to the case of multi-cell. In a multi-cell scenario, other than intra-cluster interference and inter-cluster interference, the users would suffer from inter-cell interference, which is given by

$$I_{4,m,n} = \sum_{l=1}^{L} \sum_{j=1}^{M} \sum_{i=1}^{N_j} \sqrt{\beta_{0,m,n}^{l,j,i} V^{l,j,i}} \left(g_{0,m,n}^{l,j,i} \right)^H v^{l,j,i} d^{l,j,i}, \tag{4.37}$$

where $I_{4,m,n}$ is the received inter-cell interference at the $UE_{m,n}$ of the current cell, L is the number of the interfering cells, $\beta_{0,m,n}^{l,j,i}$ and $g_{0,m,n}^{l,j,i}$ are respectively the path loss and small-scale fading of the interference channel, $d^{l,j,i}$, $V^{l,j,i}$, and $v^{l,j,i}$ are respectively the transmit signal, transmit power, and transmit beam for the $UE_{j,i}$ of the lth interfering cell. Thus, the power of the inter-cell interference can be computed as

$$\phi_{4,m,n} = \sum_{l=1}^{L} \sum_{j=1}^{M} \sum_{i=1}^{N_j} \beta_{0,m,n}^{l,j,i} V^{l,j,i} E\left[\left| \left(g_{0,m,n}^{l,j,i} \right)^H v^{l,j,i} \right|^2 \right]. \tag{4.38}$$

If there is no cooperation between the cells, the beam $\mathbf{v}^{l,j,i}$ is independent of the interference channel $\mathbf{g}_{0,m,n}^{l,j,i}$ and thus we have $\mathrm{E}\left[\left|\left(\mathbf{g}_{0,m,n}^{l,j,i}\right)^{H}\mathbf{v}^{l,j,i}\right|^{2}\right]=1$. Therefore, the power of the inter-cell interference is reduced to

$$\phi_{4,m,n}=\sum_{l=1}^{L}\sum_{j=1}^{M}\sum_{i=1}^{N_j}\beta_{0,m,n}^{l,j,i}V^{l,j,i}. \tag{4.39}$$

Adding the inter-cell interference (4.39) into the denominator of (4.36) yields the spectral efficiency in the multi-cell scenario. It is clear that the inter-cell interference leads to a reduction of spectral efficiency. In order to improve the spectral efficiency, it is desired to design the transmit beams according to the interfering CSI. However, such a beam design scheme might raise the following issues:

1. It requires the exchange of CSI for multi-cell cooperation leading to extra signalling overhead;
2. It increases the computational complexity for the design of transmit beams.
3. It consumes partial degrees of freedom for interference cancellation and thus the degrees of freedom for information transmission reduces.

In fact, as will be analyzed in the next section, if the number of BS antennas N_t is sufficiently large, the impact of inter-cell interference $\phi_{4,m,n}$ on the system performance is negligible. Since massive MIMO is a candidate technique of future 5G mobile communication systems [8], it is likely to replace the use of coordinated beamforming between multiple cells with the deployment of a large-scale antenna array at the BS to mitigate the inter-cell interference.

4.3.3   Asymptotic Characteristics of Fully Non-orthogonal Communication

In last section, we have derived a closed-form expression for a lower bound on the spectral efficiency. In order to reveal important insights of the impacts of system parameters on the spectral efficiency, we carry out asymptotic analysis on the spectral efficiency in the cases of a large number of BS antennas, a high BS transmit power, and perfect CSI at the BS. In what follows, we discuss the three important cases in detail.

4.3.3.1   A Large Number of BS Antennas

As mentioned in the system model, we consider that the BS is equipped with a large-scale antenna array. Usually, the number of antennas N_t of a large-scale antenna array in future 5G systems is more than 64. If N_t is sufficiently large, the asymptotic spectral efficiency can be expressed as

$$
\underline{R}_{m,n}^{\text{Asym1}} = \lim_{N_t \to \infty} \underline{R}_{m,n}
$$

$$
= \log_2 \left(1 + \frac{\displaystyle \lim_{N_t \to \infty} \frac{\phi_{0,m,n}}{N_t}}{\displaystyle \sum_{l=1}^{3} \lim_{N_t \to \infty} \frac{\phi_{l,m,n}}{N_t} + \lim_{N_t \to \infty} \frac{1}{N_t}} \right). \tag{4.40}
$$

We first analyze the asymptotic characteristic of the desired signal. According to the expression for the power of the desired signal in (4.29), we have

$$
\lim_{N_t \to \infty} \frac{\phi_{0,m,n}}{N_t} = \alpha_{m,n} P_{m,n} \rho_{m,n} \lim_{N_t \to \infty} \frac{\frac{\Gamma^2(N_t+1/2)}{\Gamma^2(N_t)}}{N_t}
$$

$$
= \alpha_{m,n} P_{m,n} \rho_{m,n}, \tag{4.41}
$$

where Eq. (4.41) holds since $\frac{\Gamma^2(N_t+1/2)}{\Gamma^2(N_t)}$ has the same scaling order as $N_t \to \infty$. Similarly, for the asymptotic characteristics of the powers of the signal leakage and the intra-cluster interference, we have

$$
\lim_{N_t \to \infty} \frac{\phi_{1,m,n}}{N_t} = \alpha_{m,n} P_{m,n} \rho_{m,n} \left[\lim_{N_t \to \infty} \left(\frac{N_t - 1}{N_t} - \frac{\Gamma^2(N_t + 1/2)}{N_t \Gamma^2(N_t)} \right) \right] = 0, \tag{4.42}
$$

and

$$
\lim_{N_t \to \infty} \frac{\phi_{2,m,n}}{N_t} = \alpha_{m,n} \rho_{m,n} \sum_{i=1,i \neq n}^{N_m} \kappa_{m,i} P_{m,i}, \tag{4.43}
$$

respectively. Moreover, since the powers of the inter-cluster interference and the noise are independent of N_t, we obtain

$$
\lim_{N_t \to \infty} \frac{\phi_{3,m,n}}{N_t} = 0, \tag{4.44}
$$

and

$$
\lim_{N_t \to \infty} \frac{1}{N_t} = 0, \tag{4.45}
$$

respectively.

It is found that for a sufficiently large number of BS antennas, the negative impact of the signal leakage, the inter-cluster interference, and the noise on the system performance are negligible. Hence, the asymptotic spectral efficiency can be simplified as

$$
\underline{R}_{m,n}^{\mathrm{Asym1}} = \log_2 \left(1 + \frac{\alpha_{m,n}\, P_{m,n}\, \rho_{m,n}}{\alpha_{m,n}\, \rho_{m,n} \sum\limits_{i=1,i\neq n}^{N_m} \kappa_{m,i}\, P_{m,i}} \right)
$$

$$
= \log_2 \left(1 + \frac{P_{m,n}}{\sum\limits_{i=1,i\neq n}^{N_m} \kappa_{m,i}\, P_{m,i}} \right). \tag{4.46}
$$

Interestingly, it is seen that the spectral efficiency in the case of a large number of BS antennas has the following asymptotic characteristics:

1. It is independent of path loss, since the desired signal and the intra-cluster interference experience the same channel fading.
2. It is independent of CSI accuracy as long as the BS has partial CSI, i.e., $\rho_{m,n} > 0$, since the desired signal and the intra-cluster interference are projected into the same space.
3. It is independent of the parameters of other clusters, since the inter-cluster interference is negligible.

In summary, if the number of BS antennas is sufficiently large, very high-resolution beams can be formed for carrying information. Under this condition, the interference from different beams, namely the inter-cluster interference, is negligible. However, in a cluster, the signals of different users are treated in the same ways, such that the intra-cluster interference always exists. The intra-cluster interference is mainly determined by transmit powers and residual interference factors. Hence, the transmit power optimization is necessary for improving the spectral efficiency. In addition, it is worth pointing out that if orthogonal transmission is adopted, namely $N_m = 1, \forall m$, the intra-cluster interference does not exist. However, in practical systems, N_t is not so large such that it is not able for orthogonal transmission to admit massive connections. Moreover, orthogonal transmission suffers from the strong inter-cluster interference, while non-orthogonal transmission is capable of achieving a balance between the inter-cluster interference and the intra-cluster interference.

In the multi-cell scenario, the inter-cell interference in (4.39) is also independent of N_t. As a result, the inter-cell interference is negligible as N_t asymptotically approaches infinity, which reconfirms the claim at the end of Sect. 4.3.2. Therefore, the asymptotic spectral efficiencies in a single-cell scenario and a multi-cell scenario are the same.

4.3.3.2   High BS Transmit Power

In this subsection, we analyze the asymptotic characteristics of the spectral efficiency for a high BS transmit power. For ease of analysis, we let $P_{m,n} = \mu_{m,n} P$ with $\sum_{m=1}^{M} \sum_{n=1}^{N_m} \mu_{m,n} = 1$, where P is the total transmit power of the BS, and $\mu_{m,n}$ is the coefficient of power allocation for the $UE_{m,n}$. Thus, the asymptotic spectral efficiency at a high BS transmit power can be expressed as

$$
\underline{R}_{m,n}^{\mathrm{Asym2}} = \lim_{P_{tot} \to \infty} \underline{R}_{m,n}
$$

$$
= \log_2 \left(1 + \frac{\lim\limits_{P \to \infty} \frac{\Phi_{0,m,n}}{P}}{\sum\limits_{t=1}^{3} \lim\limits_{P \to \infty} \frac{\Phi_{t,m,n}}{P} + \lim\limits_{P \to \infty} \frac{1}{P}} \right) \approx \log_2 \left(1 + \frac{\lim\limits_{P \to \infty} \frac{\Phi_{0,m,n}}{P}}{\sum\limits_{t=1}^{3} \lim\limits_{P \to \infty} \frac{\Phi_{t,m,n}}{P}} \right) \quad (4.47)
$$

$$
= \log_2 \left(1 + \frac{\mu_{m,n} \rho_{m,n} \frac{\Gamma^2(N_t + 1/2)}{\Gamma^2(N_t)}}{\left[\left(\rho_{m,n}(N_t - 1) + 1 \right) \sum\limits_{i=1}^{N_m} \kappa_{m,i} \mu_{m,i} \right.} \right),
$$
$$
\left. {} - \mu_{m,n} \rho_{m,n} \frac{\Gamma^2(N_t + 1/2)}{\Gamma^2(N_t)} + \sum\limits_{j=1, j \neq m}^{M} \sum\limits_{i=1}^{N_j} \mu_{j,i} \right]
$$

$$(4.48)$$

where Eq. (4.47) holds true because the noise term is negligible at the high transmit power regime. Note that the asymptotic spectral efficiency $\underline{R}_{m,n}^{\mathrm{Asym2}}$ is independent of P. In other words, for a given power allocation scheme, the performance is saturated if the total transmit power is large enough. In this case, it is possible to enhance the performance by improving the accuracy of CSI estimation. Moreover, it is found that the performance upper bound is regardless of path loss. This is because all the signals experience the same channel fading.

In the multi-cell scenario, the performance ceiling also appears at a high transmit power, if the BSs have the same order of transmit power.

4.3.3.3   Perfect CSI at the BS

In some special cases, the BS may obtain very accurate CSI for a certain user, namely $\rho_{m,n} \to 1$. Hence, the asymptotic spectral efficiency in the presence of accurate CSI can be computed as

$$\underline{R}_{m,n}^{\text{Asym3}} = \lim_{\rho_{m,n} \to 1} \underline{R}_{m,n}$$

$$= \log_2 \left(1 + \frac{\alpha_{m,n} P_{m,n} \frac{\Gamma^2(N_t+1/2)}{\Gamma^2(N_t)}}{\alpha_{m,n} N_t \sum_{i=1}^{N_m} \kappa_{m,i} P_{m,i} - \alpha_{m,n} P_{m,n} \frac{\Gamma^2(N_t+1/2)}{\Gamma^2(N_t)} + \sum_{j=1,j\neq m}^{M} \sum_{i=1}^{N_j} \alpha_{m,n} P_{j,i} + 1} \right).$$

$$\tag{4.49}$$

Given the number of clusters M and the number of users in a cluster N, $\underline{R}_{m,n}^{\text{Asym3}}$ in (4.49) is the maximum value of the lower bound on the spectral efficiency. Moreover, it is noted that the desired signal, the signal leakage and the residual intra-cluster interference are a function of the CSI accuracy. Hence, it makes sense to select an optimal N_m according to the CSI accuracy, so as to maximize the spectral efficiency.

In the multi-cell scenario, if there is no multi-cell cooperation, the improvement of CSI estimation accuracy does not reduce the inter-cell interference. In contrast, when the BS are fully cooperative, the accurate CSI can be used to mitigate the inter-cell interference. However, coordinated beamforming for cancelling the inter-cell interference consumes a partial of degrees of freedom, which may decrease the quality of the desired signal. Thus, it makes sense to achieve a tradeoff between enhancing the desired signal and mitigating the interference.

4.4 Optimization of Fully Non-orthogonal Communication

As can be seen in (4.36), the spectral efficiency is determined by the parameters in both channel estimation and multiple access, i.e., transmit power. Thus, it is likely to further improve the performance by optimizing these system parameters. In this section, we propose two simple but effective algorithms to optimize channel estimation and multiple access for maximizing the weighted sum of spectral efficiency. Moreover, we also provide the optimization methods based on other performance metrics, e.g., total power consumption, energy efficiency, and user fairness.

4.4.1 Optimization of Non-orthogonal Channel Estimation

It is intuitive that the CSI obtained by channel estimation has a great impact on the spectral efficiency. For the proposed non-orthogonal channel estimation, the CSI accuracy of an arbitrary user is determined by the transmit powers of the users for a given length of pilot sequences. Hence, the optimization of channel estimation is equivalent to controlling the transmit powers of the users.

Since pilot sequences are orthogonal among different clusters and the inter-cluster interference is independent of the CSI accuracy, it is likely to optimize the transmit powers of the users in each cluster independently. We take the optimization of the channel estimation for the mth cluster as an example. Note that given the length of pilot sequence τ and path loss $\alpha_{m,i}$, the CSI accuracy $\rho_{m,i}$ and transmit power $Q_{m,i}$ have an one-to-one mapping relation. In this case, we replace $Q_{m,i}$ with $\rho_{m,i}$ as the optimization variable. Then, the maximization of the weighted sum of spectral efficiency can be described as the following optimization problem:

$$\text{OP1} : \max_{\rho_m} \sum_{i=1}^{N} \theta_{m,i} \underline{R}_{m,i}$$

$$\text{s.t. C1} : \rho_{m,n} \geq 0, \forall n$$

$$\text{C2} : \sum_{i=1}^{N_m} \rho_{m,i} \leq 1 - \frac{1}{\sum_{i=1}^{N_m} \alpha_{m,i} Q_{m,i}^{\max} \tau}, \tag{4.50}$$

where $Q_{m,i}^{\max}$ is the maximum available transmit power of the ith user in the mth cluster and ρ_m is a collection of $\rho_{m,i}, \forall i$. The positive constant $\theta_{m,i}$ denotes the priority such that $\sum_{m=1}^{M} \sum_{i=1}^{N_m} \theta_{m,i} = 1$. Note that $\underline{R}_{m,i}$ is a concave function of $\rho_{m,i}$, thus OP1 can be solved by the Lagrange multiplier method [23]. First, we construct the Lagrange function of OP1 as follows:

$$\mathscr{L}(\rho_m) = -\sum_{i=1}^{N} (\theta_{m,i} \underline{R}_{m,i} + \nu_{m,i} \rho_{m,i}) + \varphi_m \left(\sum_{i=1}^{N_m} \rho_{m,i} - \left(1 - \frac{1}{\sum_{i=1}^{N_m} \alpha_{m,i} Q_{m,i}^{\max} \tau} \right) \right), \tag{4.51}$$

where $\nu_{m,i} \geq 0$ and $\varphi_m \geq 0$ are the Lagrange multipliers of C1 and C2, respectively. Then, by exploiting the Karush-Kuhn-Tucker (KKT) conditions [23], we have

$$\frac{\partial \mathscr{L}(\rho_m)}{\partial \rho_{m,n}} = -\frac{\theta_{m,n}}{\ln 2} \frac{c_{m,n,1} c_{m,n,3}}{((c_{m,n,1} + c_{m,n,2})\rho_{m,n} + c_{m,n,3})(c_{m,n,2}\rho_{m,n} + c_{m,n,3})}$$
$$-\nu_{m,n} + \varphi_m = 0. \tag{4.52}$$

where $c_{m,n,1} = \alpha_{m,n} P_{m,n} \frac{\Gamma^2(N_t+1/2)}{\Gamma^2(N_t)}$, $c_{m,n,2} = \alpha_{m,n}(N_t - 1) \sum\limits_{i=1}^{N_m} \kappa_{m,i} P_{m,i} -$

$\alpha_{m,n} P_{m,n} \frac{\Gamma^2(N_t+1/2)}{\Gamma^2(N_t)}$, and $c_{m,n,3} = \alpha_{m,n} \sum\limits_{i=1}^{N_m} \kappa_{m,i} P_{m,i} + \alpha_{m,n} \sum\limits_{j=1, j \neq m}^{M} \sum\limits_{i=1}^{N_j} P_{j,i} + 1$.

By solving Eq. (4.52), we can obtain $\rho_{m,n}$ as

$$\rho_{m,n} = \frac{\left[-(c_{m,n,1}c_{m,n,3} + 2c_{m,n,2}c_{m,n,3}) + \sqrt{c_{m,n,2}^2 c_{m,n,3}^2 + \frac{4c_{m,n,1}c_{m,n,2}c_{m,n,3}(c_{m,n,1}+c_{m,n,2})\theta_{m,n}}{(\varphi_m - \nu_{m,n})\ln 2}} \right]}{2(c_{m,n,1} + c_{m,n,2})c_{m,n,3}}. \tag{4.53}$$

By iteratively updating the multipliers $\nu_{m,i}$ and φ_m with the gradient method [23], one can derive the solution of $\rho_{m,n}$. In summary, the proposed power control algorithm for non-orthogonal channel estimation can be described as

---

Algorithm 1 Power control for non-orthogonal channel estimation

---

Step 1: Initialize the parameters by letting $Q_{m,n} = Q_{m,n}^{\max}$ and $\rho_{m,n} = \frac{\alpha_{m,n} Q_{m,n}\tau}{1+ \sum\limits_{j=1}^{N_m} \alpha_{m,j} Q_{m,j}\tau}$, $\forall m, n$;

Step 2: Set $\nu_{m,n} = 1$ and $\varphi_m = 1$;
Step 3: Update $\rho_{m,n}$ according to (4.53), and update $\nu_{m,n}$ and φ_m by the gradient method. Redo Step 3 until $\nu_{m,n}$ and φ_m are converged;
Step 4: Go to Step 2 until the weighted sum of spectral efficiency is converged;
Step 5: Compute and output $Q_{m,n}$ according to the relation between $\rho_{m,n}$ and $Q_{m,n}$.

Note that since OP1 is a convex problem, the transmit powers at the users can be derived in polynomial time.

4.4.2 Optimization of Non-orthogonal Multiple Access

As analyzed earlier, the performance of channel estimation is only related to the users in a cluster, but the design of multiple access involves all users. Thus, it is possible to achieve higher performance gains by optimizing the power allocation in non-orthogonal multiple access.

The optimization of non-orthogonal multiple access for maximizing the weighted sum of spectral efficiency is equivalent to the following optimization problem:

$$\text{OP2} : \max_{\mathbf{P}} \sum_{j=1}^{M} \sum_{i=1}^{N} \theta_{j,i} \underline{R}_{j,i}$$

$$\text{s.t. C3} : P_{m,n} \geq 0, \forall m, n,$$

$$C4 : \sum_{j=1}^{M} \sum_{i=1}^{N_j} P_{j,i} \leq P_{\text{tot}}, \tag{4.54}$$

where $\mathbf{P}$ is a collection of the transmit powers $P_{m,n}$ and P_{tot} is the maximum power budget at the BS. To solve OP2, the key is to transform the objective function to a concave function. Since the objective function is a weighted sum of multiple logarithmic functions, we can utilize the sequential convex approximation (SCA) method to approximate the objective function by a convex function [23]. Subsequently, we iteratively optimize an approximated convex problem until the solution converges. In each iteration, we leverage the following lower bound on the logarithmic function to approximate the objective function:

$$\log_2(1 + \gamma) \geq a \log_2 \gamma + b, \tag{4.55}$$

where $a = \frac{\tilde{\gamma}}{1+\tilde{\gamma}}$ and $b = \log_2(1 + \tilde{\gamma}) - \frac{\tilde{\gamma}}{1+\tilde{\gamma}} \log_2(\tilde{\gamma})$ with $\tilde{\gamma}$ being the SINR obtained at the last iteration. Thus, the term $\underline{R}_{m,n}$ in the objective function can be approximated as

$$\underline{R}_{m,n} \geq a_{m,n} \log_2(\underline{\gamma}_{m,n}) + b_{m,n}$$

$$= a_{m,n} \log_2 \left(\alpha_{m,n} \rho_{m,n} \frac{\Gamma^2(N_t + 1/2)}{\Gamma^2(N_t)} P_{m,n} \right)$$

$$- a_{m,n} \log_2 \left(\sum_{j=1}^{M} \sum_{i=1}^{N_j} \varphi_{j,i} P_{j,i} + 1 \right) + b_{m,n}$$

$$= a_{m,n} \left(\log_2 \left(\alpha_{m,n} \rho_{m,n} \frac{\Gamma^2(N_t + 1/2)}{\Gamma^2(N_t)} \right) + P'_{m,n} \right)$$

$$- a_{m,n} \log_2 \left(\sum_{j=1}^{M} \sum_{i=1}^{N_j} \varphi_{j,i} 2^{P_{j,i}} + 1 \right) + b_{m,n}, \tag{4.56}$$

where $a_{m,n} = \frac{\tilde{\gamma}_{m,n}}{1+\tilde{\gamma}_{m,n}}$, $b_{m,n} = \log_2(1 + \underline{\tilde{\gamma}}_{m,n}) - \frac{\tilde{\gamma}_{m,n}}{1+\underline{\tilde{\gamma}}_{m,n}} \log_2(\underline{\tilde{\gamma}}_{m,n})$, $P'_{j,i} = \log_2(P_{j,i})$, and

$$\varphi_{j,i} = \begin{cases} \alpha_{m,n} \left(\rho_{m,n} \left(N_t - 1 - \frac{\Gamma^2(N_t+1/2)}{\Gamma^2(N_t)} \right) + 1 \right), & \text{if } j = m \text{ and } i = n, \\ \alpha_{m,n}(\rho_{m,n}(N_t - 1) + 1)\kappa_{m,i}, & \text{if } j = m \text{ and } i \neq n, \\ \alpha_{m,n}, & \text{otherwise.} \end{cases}$$

Since the log-sum-exp function is convex [23], Eq. (4.56) is a concave function. Thus, the objective function can be approximated as a weighed sum of concave functions which is also a concave function. Then, we can solve OP2 by iteratively updating $\tilde{\gamma}_{m,n}$, $\forall m, n$. In each iteration, the optimization problem is convex which can be solved by some optimization softwares, such as CVX [22]. Thus, the proposed power allocation algorithm for non-orthogonal multiple access can be summarized as

---

Algorithm 2 Power allocation for non-orthogonal multiple access

---

Step 1:   Initialize the parameters by setting $P_{m,n} = \frac{P_{\text{tot}}}{K}$, $\forall m, n$;

Step 2:   Update $\tilde{\gamma}_{m,n} = \alpha_{m,n} P_{m,n} \rho_{m,n} \frac{\Gamma^2(N_t+1/2)}{\Gamma^2(N_t)} \Big/ \Big(\alpha_{m,n}(\rho_{m,n}(N_t - 1) + 1) \sum_{i=1}^{N_m} \kappa_{m,i} P_{m,i} - $

$\alpha_{m,n} P_{m,n} \rho_{m,n} \frac{\Gamma^2(N_t+1/2)}{\Gamma^2(N_t)} + \alpha_{m,n} \sum_{j=1, j\neq m}^{M} \sum_{i=1}^{N_j} P_{j,i} + 1 \Big)$, $\alpha_{m,n} = \frac{\tilde{\gamma}_{m,n}}{1+\tilde{\gamma}_{m,n}}$, $b_{m,n} = \log(1+\underline{\tilde{\gamma}}_{m,n}) -$

$\frac{\tilde{\gamma}_{m,n}}{1+\tilde{\gamma}_{m,n}} \log(\underline{\tilde{\gamma}}_{m,n})$, and $P'_{m,n} = \log_2(P_{m,n})$;

Step 3:   Obtain $P'_{m,n}$ by solving the approximated OP2 using CVX;

Step 4:   Go to Step 2 until $P'_{m,n}$ is converged;

Step 5:   Output $P_{m,n} = 2^{P'_{m,n}}$.

Since we use an approximated objective function, Algorithm 2 can only achieve a suboptimal performance. Moreover, according to the property of the SCA method [23], the weighted sum spectral efficiency is increased after each iteration. Thus, Algorithm 2 always converges due to the existence of the upper bound of the weighted sum of spectral efficiency. It is clear that the computational complexity of Algorithm 2 is determined by the number of iterations. In other words, for a given maximum number of iterations, Algorithm 2 has a polynomial time computational complexity.

Remark 4 In this section, we provide a framework for the optimization of channel estimation and multiple access from the perspective of maximizing the weighted sum of spectral efficiency. It is worth pointing out that the proposed framework can be easily extended to the optimization with the other important performance metrics, e.g., total power consumption, energy efficiency, and user fairness. In particular, for prolonging the lifetime of wireless networks, it is desired to minimize the total power consumption subject to a constraint on the minimum spectral efficiency of each user. Such a problem can be transformed as a linear programming problem, which is easily solved by off-the-shelf optimization softwares. Moreover, it is likely to design an energy-efficiency power allocation scheme by maximizing the energy efficiency with the fractional programming method [24], and a user fairness power allocation scheme by maximizing the minimum spectral efficiency through iteratively updating the minimum spectral efficiency [25].

4.5 Numerical Results

In this section, we conduct extensive simulations to validate the effectiveness of the proposed fully non-orthogonal communication for massive access with imperfect SIC in a single-cell cellular IoT. For convenience, we set the simulation parameters as follows: $N_t = 64$, $K = 48$, $M = 12$, $N_m = N = 4$, $\eta_{m,n} = \eta = 0.1, \forall m, n$. Moreover, we assume that the normalized path losses of the channels for all UEs form an arithmetic sequences from 1 to 0.04 with a common difference -0.02. We use the reminder of dividing the user index by M to determine user clustering. For instance, the kth user belongs to the $\mathrm{mod}(k/M)$th cluster. If the reminder equals to zero, the corresponding user is in the Mth cluster. Through allocating transmit powers of pilot sequences, we let all CSI correlation coefficients $\rho_{m,n}$ equal to $\rho = 0.24$. Moreover, it is assumed that the transmit power P of the BS is equally distributed to the UEs for information transmission, and we use SNR $= 10 \log_{10} P$ to denote the transmit SNR (in dB). Without extra specification, we focus on the spectral efficiency of the 1st UE in the 1st cluster.

Firstly, we verify the accuracy of the derived theoretical expression for a lower bound on the spectral efficiency with different numbers of BS antennas. As seen in Fig. 4.2, the gaps between the simulation results and the theoretical lower bound are negligible when the total transmit SNR is larger than 10 dB (-6.8 dB for each user's signal). Thus, the derived theoretical lower bound is tight in practical SNR region. Moreover, it is found that the spectral efficiency is saturated in the high SNR region, which reconfirms the conclusion based on Eq. (4.48). Moreover, given a transmit

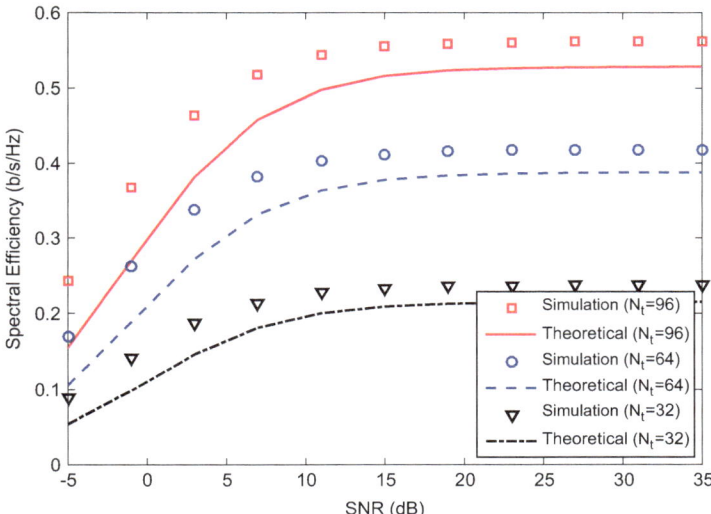

Fig. 4.2 Comparison of theoretical expressions and simulation results

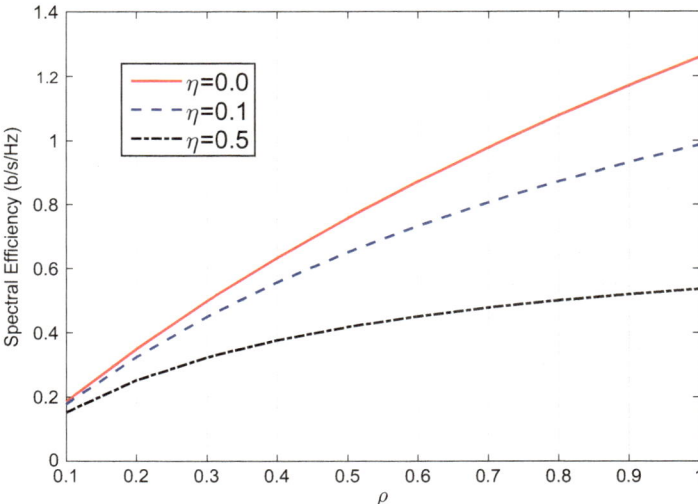

Fig. 4.3 Impact of CSI accuracy and SIC residual interference factor

SNR, the performance ceiling can be improved by adding more antennas but with diminishing returns.

Secondly, we investigate the impact of CSI accuracy and SIC residual interference factor on the spectral efficiency at SNR = 15 dB, cf. Fig. 4.3. It is seen that if the CSI accuracy is low, the impact of SIC residual interference factor is insignificant. This is because the power of the residual interference caused by imperfect SIC is a linear function of the CSI correlation coefficient ρ. As mentioned earlier, the CSI accuracy in the context of non-orthogonal channel estimation is limited by the number of users in a cluster, resulting in its insensitivity to imperfect SIC. For instance, if there are $N = 4$ users in a cluster, the average CSI correlation coefficient is less than $\rho = 0.25$. Under this condition, the performance loss caused by $\eta = 0.1$ is negligible. Hence, an extra advantage of the proposed fully non-orthogonal communication is the robustness against the imperfectness of SIC.

Thirdly, we compare the performance of two commonly used beamforming schemes based on the proposed full non-orthogonal communication framework, i.e., MF and ZF. As can be seen from Fig. 4.4, the spectral efficiency of the both beamforming schemes would be saturated in a high SNR region. With $\rho = 0.24$, MF beafmorming scheme outperforms ZF beamforming significantly and the performance gain is enlarged as SNR increases. This is because ZF beamforming scheme requires more spatial degrees of freedom to mitigate the inter-cluster interference. If the CSI accuracy is low, the capability of ZF beamforming scheme is limited. However, with the increase of the CSI accuracy, the performance of ZF beamforming scheme might be better than that of MF beamformign scheme. Therefore, it makes sense to select a proper beamforming scheme adaptively according to system parameters and channel conditions.

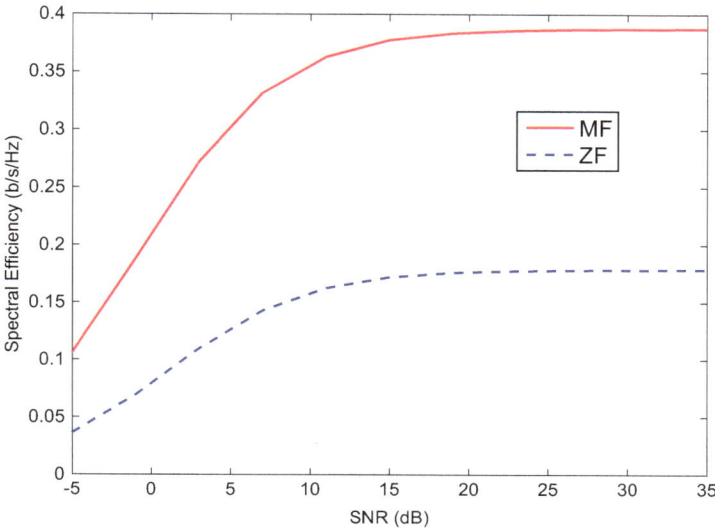

Fig. 4.4 Performance comparison of MF beamforming and ZF beamforming

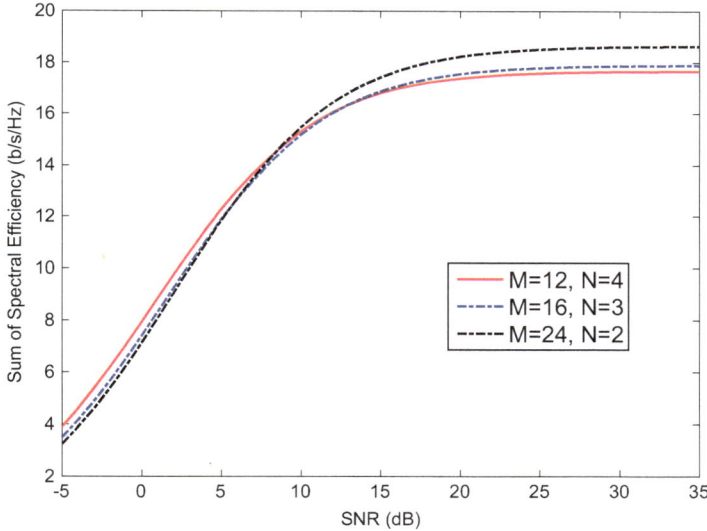

Fig. 4.5 Performance comparison of different user clustering modes

Then, we examine the effect of user clustering mode on the sum of spectral efficiency. For a given number of users $K = 48$, we select three typical user clustering modes, i.e., "$M = 12, N = 4$", "$M = 16, N = 3$", and "$M = 24, N = 2$". As shown in Fig. 4.5, in a low SNR region, the first mode performs best. This is because the inter-cluster interference is dominant in this scenario, and the first mode may suffer from the lowest inter-cluster interference. As SNR increases, the third mode achieves the best performance, since this scenario is the intra-cluster interference limited. It is worth noting that M and N determine the number of beams and the order of SIC. Thus, we can dynamically adjust the user clustering mode according to system parameters and channel conditions for balancing the overall performance and the computational complexity.

Next, we compare the performance of three non-orthogonal channel estimation optimization schemes, i.e., proposed power control, equal power control, and equal accuracy power control. As the names implies, the three schemes determine the powers of pilot sequences by solving OP1 with equal weighted coefficients, equal distribution, and making the BS to have the same CSI accuracies of all the downlink channels, respectively. Equal power control and equal accuracy power control avoid the iterations for computing the transmit powers, thus have lower complexity. The complexity of the proposed power control scheme is determined by the number of iterations. In simulations, it is found that the proposed power control scheme converges within five iterations, thus its complexity is bearable in practical systems. In terms of sum spectral efficiency, as seen in Fig. 4.6, the proposed power control scheme performs the best in the whole SNR region, since it is designed from the perspective of maximizing the sum of spectral efficiency. The equal power control scheme nearly achieves the same performance as the optimal one. This is because this scheme forces the user with the largest channel variance to obtain the most accurate CSI, which is beneficial to improve the sum of spectral efficiency. Thus,

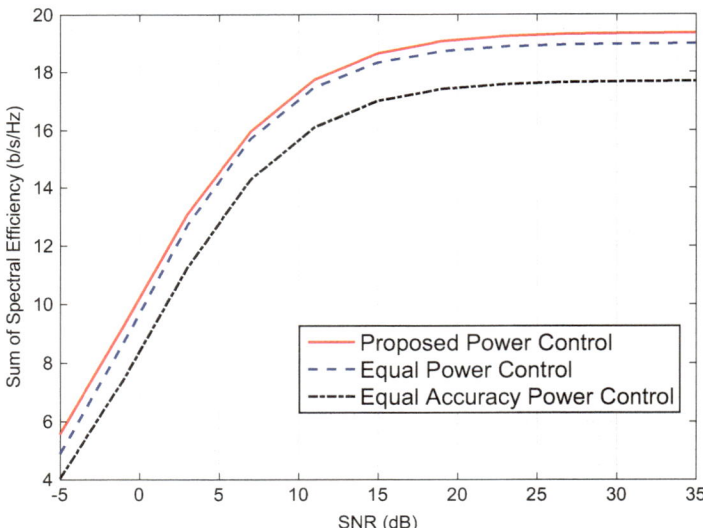

Fig. 4.6 Performance comparison of different optimization schemes for non-orthogonal channel estimation

it is appealing to adopt the equal power control scheme to obtain the near-optimal performance with a low complexity. However, the equal power control scheme is not able to guarantee user fairness. On the other hand, the equal accuracy power control scheme can improve the minimum spectral efficiency, although it might lead to a loss in the sense of the sum of spectral efficiency. Hence, it makes sense to choose a proper non-orthogonal channel estimation optimization scheme according to the characteristics of wireless services.

Finally, we show the performance gain of the proposed power allocation scheme for non-orthogonal multiple access over two baseline schemes, namely equal power allocation and fixed power allocation. Note that fixed power scheme proposed in [26] distributes $\frac{n}{\sum_{i=1}^{N} i} \frac{P_{tot}}{M}$ to the nth user in each cluster fixedly. In simulations, the proposed power allocation scheme converges within 10 iterations and thus it has bearable complexity. As seen in Fig. 4.7, the proposed power allocation scheme performs much better than the two baseline schemes in the whole SINR region, since it adjusts the power allocation according to channel conditions and system parameters. Moreover, it is found that the performance of fixed power allocation scheme is even worse than that of equal power allocation. This is because fixed allocation scheme distributes more power to the user with a weak channel gain, resulting in a lower power utilization efficiency.

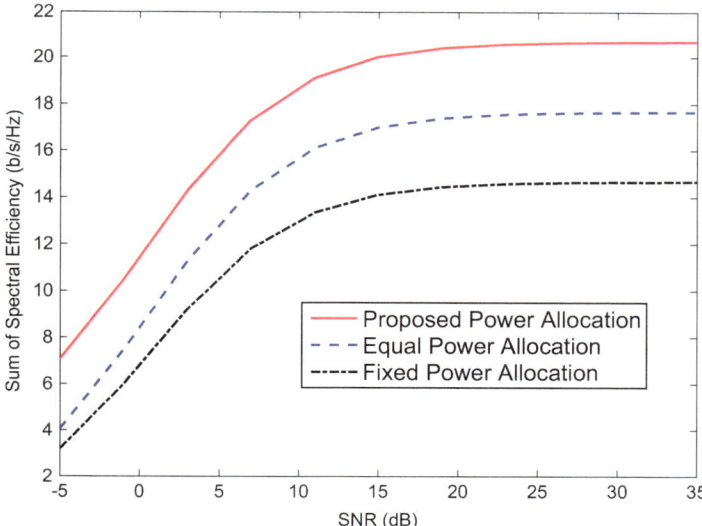

Fig. 4.7 Performance comparison of different optimization schemes for non-orthogonal multiple access

4.6   Conclusion

In this chapter, we have designed a fully non-orthogonal communication framework by combining non-orthogonal channel estimation and non-orthogonal multiple access to support massive access over limited radio spectrum. In particular, non-orthogonal channel estimation was proposed to solve the practical issue of CSI acquisition. Besides, non-orthogonal multiple access was optimized to alleviate the impact of imperfect SIC. Both theoretical analysis and simulation results showed that the proposed fully non-orthogonal communication can effectively improve the spectral efficiency and thus enable massive access in future 5G cellular IoT.

References

1. M. Shirvanimoghaddam, M. Dohler, S.J. Johnson, Massive non-orthogonal multiple access for cellular IoT: potentials and limitations. IEEE Commun. Mag. **55**(9), 55–61 (2017)
2. X. Chen, Z. Zhang, C. Zhong, R. Jia, D.W.K. Ng, Fully non-orthogonal communication for massive access. IEEE Trans. Commun. **66**(4), 1717–1731 (2018)
3. O. Elijah, C.Y. Leow, T.A. Rahman, S. Nunoo, S.Z. Lliya, A comprehensive survey of pilot contamination in massive MIMO-5G system. IEEE Commun. Surv. Tuts **18**(2), 905–923 (2016)
4. X. Chen, Z. Zhang, C. Zhong, D.W.K. Ng, R. Jia, Exploiting inter-user interference for secure massive non-Orthogonal multiple access. IEEE J. Sel. Areas Commun. **36**(4), 788–801 (2018)
5. A. Bayesteh, A.K. Khandani, Asymptotic analysis of the amount of CSI feedback in MIMO broadcast channels. IEEE Trans. Inf. Theory **58**(3), 1612–1629 (2012)
6. N. Jindal, MIMO broadcast channels with finite-rate feedback. IEEE Trans. Inf. Theory **52**(11), 5045–5060 (2006)
7. T.L. Marzetta, Noncooperative celluar wireless with unlimited numbers of base station antennas. IEEE Trans. Wirel. Commun. **9**(11), 3590–3600 (2010)
8. J. Hoydis, S.T. Brink, M. Debbah, Massive MIMO in the UL/DL of cellular networks: how many antennas do we need. IEEE J. Sel. Areas Commun. **31**(2), 160–171 (2013)
9. E.G. Larsson, H.V. Poor, Joint beamforming and broadcasting in massive MIMO. IEEE Trans. Wirel. Commun. **15**(4), 3058–3070 (2016)
10. J. Jose, A. Ashikhmin, T.L. Marzetta, S. Vishwanath, Pilot contamination and precoding in multi-cell TDD systems. IEEE Trans. Wirel. Commun. **10**(8), 2640–2651 (2011)
11. J. Zhang, R. Schober, V.K. Bhargava, Secure transmission in multicell massive MIMO systems. IEEE Trans. Wirel. Commun. **13**(9), 4766–4781 (2014)
12. T.V. Chien, E. Bjornson, E.G. Larsson, Joint power allocation and user association optimization for massive MIMO systems. IEEE Trans. Wirel. Commun. **15**(9), 6384–399 (2016)
13. M.F. Hanif, Z. Ding, T. Ratnarajah, G.K. Karagiannidis, A minorization-maximization method for optimizing sum rate in the downlink of non-orthogonal multiple access systems. IEEE Trans. Signal Process. **64**(1), 76–88 (2016)
14. C. Zhong, Z. Zhang, Non-orthogonal multiple access with cooperative full-duplex relaying. IEEE Commun. Lett. **20**(12), 2478–2481 (2016)
15. W. Shin, M. Vaezi, B. Lee, D.J. Love, J. Lee, H.V. Poor, Coordinated beamforming for multi-cell MIMO-NOMA. IEEE Commun. Lett. **21**(1), 84–87 (2017)
16. S.M.R. Islam, N. Avazov, O.A. Dobre, K.-S. Kwak, Power-domain non-orthogonal multiple access (NOMA) in 5G: potentials and challenges. IEEE Commun. Surv. Tuts **19**(2), 721–742 (2017)

17. J.G. Andrews, S. Buzzi, W. Choi, S.V. Hanly, A. Lozano, A.C.K. Soong, J.C. Zhang, What will 5G be? IEEE J. Sel. Areas Commun. **32**(6), 1065–1082 (2014)
18. V. Raghavan, S. Subramanian, J. Cezanne, A. Sampath, O.H. Koymen, J. Li, Single-user versus multi-user precoding for millimeter wave MIMO systems. IEEE J. Sel. Areas Commun. **35**(6), 1387–1401 (2017)
19. S. Roger, D. Calabuig, J. Cabrejas, J.F. Monserrat, Multi-user non-coherent detection for downlink MIMO communication. IEEE Signal Process. Lett. **21**(10), 1225–1229 (2014)
20. V. Raghavan, J.H. Kotecha, A.M. Sayeed, Why does the Kronecker model result in misleading capacity estimates? IEEE Trans. Inf. Theory **56**(10), 4843–4864 (2010)
21. P. Li, R.C. de Lamare, R. Fa, Multiple feedback successive interference cancellation detection for multiuser MIMO systems. IEEE Trans. Wirel. Commun. **10**(8), 2434–2439 (2011)
22. M. Grant, S. Boyd, CVX: matlab software for disciplined convex programming. [Online]: http://cvxr.com/cvx
23. S. Boyd, L. Vandenberghe, *Convex Optimization* (Cambridge University Press, Cambridge, 2004)
24. X. Chen, L. Lei, Energy-efficient optimization for physical layer security in multi-antenna downlink networks with QoS guarantee. IEEE Commun. Lett. **17**(4), 637–640 (2013)
25. S. Timotheou, I. Krikidis, Fairness for non-orthogonal multiple access in 5G systems. IEEE Signal Process. Lett. **22**(10), 1647–1651 (2015)
26. Z. Ding, Z. Yang, P. Fan, H.V. Poor, On the performance of non-orthogonal multiple access in 5G systems with randomly deployed users. IEEE Signal Process. Lett. **21**(12), 1501–1505 (2014)

Chapter 5
Massive Access with Channel Statistical Information

Abstract In this chapter, we study the problem of massive access in the 5G cellular IoT, where the channels are fast-varying. To address the challenging issue of channel state information (CSI) acquisition and beam design for a massive number of IoT devices over fast time-varying fading channels, we design a non-orthogonal beamspace multiple access framework. In particular, the user equipments (UEs) are non-orthogonal not only in the temporal-frequency domain, but also in the beam domain. We analyze the performance of the proposed non-orthogonal beamspace multiple access scheme, and derive an upper bound on the weighted sum rate in terms of channel conditions and system parameters. For further improving the performance, we propose three non-orthogonal beam construction methods with different beamspace resolutions. Finally, extensively simulation results show the performance gain of the proposed non-orthogonal beamspace multiple access scheme over the baseline ones.

5.1 Introduction

To exploit the benefits of spatial degrees of freedom for user clustering and interference cancellation in the cellular IoT, the multiple-antenna BS should has channel state information about the access devices [1–3]. In the ideal case, the BS obtains full CSI by some means, and thus it is possible to perform optimal user clustering and completely cancel the inter-cluster interference [4, 5]. However, it is not trivial task to obtain CSI in the MIMO NOMA systems, since the BS is at the transmit side of the downlink channels. In practical systems, the CSI is usually obtained at the BS through quantized feedback in FDD mode [6] and channel estimation in TDD mode [7]. However, in the scenario of massive access for the cellular IoT, the resource consumption in quantized feedback or channel estimation is prohibitive. To alleviate the burden in resource consumption, a non-orthogonal channel estimation method for CSI acquisition was proposed in [8]. Due to the

X. Chen, *Massive Access for Cellular Internet of Things Theory and Technique*,
SpringerBriefs in Electrical and Computer Engineering,
https://doi.org/10.1007/978-981-13-6597-3_5

co-channel interference, the non-orthogonal channel estimation method leads to a severe reduction of the CSI accuracy compared to the orthogonal one. Moreover, the simple user clustering method based on channel spatial correlation is unfeasible and the commonly used interference cancellation method based on zero-forcing beamforming is inefficient in the case of low-precision CSI. It is worth pointing out that the CSI acquisition methods based on quantized feedback and channel estimation are applicable in the case of slow time-vary channel fading, since they require long channel coherent time [9, 10]. However, in some application scenarios of the cellular IoT, the IoT devices, e.g., vehicular equipments, may move with a high speed, resulting in fast time-varying channel fading. Then, the traditional CSI acquisition methods are in applicable. Recently, a beam division multiple access (BDMA) scheme was proposed for the multiuser massive MIMO systems [11, 12]. Specifically, the users are simultaneously served by asymptotically orthogonal beams, which are constructed only based on channel coupling matrix (namely channel correlation matrix). The orthogonal beams can effectively decrease the co-channel interference, and thus improve the overall performance. More importantly, channel coupling matrix varies over a relatively long period, and can be constructed with some large-scale parameters, e.g., angle of arrival (AoA) [13, 14]. Hence, BDMA is regarded as a simple and feasible multiple access scheme. Furthermore, if a beam of BDMA serves multiple users, namely non-orthogonal BDMA, the system is capable of supporting massive connections [15]. In other words, the clusters in the NOMA systems are separated in the beamspace, which also simplifies the user clustering [16].

For the beamspace MIMO-NOMA systems, a beam needs to serves multiple access devices in the same cluster. Due to the random distribution of the devices, a beam is unable to align all devices' channels. Therefore, although the beams are orthogonal, there still exists severe inter-cluster interference. In fact, the orthogonal beams might be not optimal, and it makes sense to redesign the beams from the perspective of improving the overall performance. In [17], the authors proposed to adjust the beam direction by padding zeros in the beam domain, so as to make the beam align the devices. However, padding zeros is an integer programming problem, which is difficult to provide a general solution. To this end, we provide a new non-orthogonal beamspace multiple access for massive connections in the cellular IoT. In particular, the access devices are non-orthogonal not only in the temporary-frequency domain, but also in beam domain. The non-orthogonal beams can be directly obtained from the originally base beams by linear combinations. The contributions of this chapter are as follows:

1. This chapter designs a comprehensive non-orthogonal beamspace multiple access scheme for the cellular IoT with massive connections, including CSI acquisition, user clustering, superposition coding and SIC.
2. This chapter analyzes the performance of the proposed non-orthogonal beamspace multiple access scheme, and derives a closed-form expression for a upper bound on the weighted sum rate in terms of system parameters and channel conditions.

3. This chapter optimizes the proposed non-orthogonal beamspace multiple access scheme, and presents three beam design algorithms with different degrees of freedom in the beamspace.

The rest of this chapter is organized as follows: Sect. 5.2 gives a brief introduction of the cellular IoT and presents a massive access framework in the beamspace. Section 5.3 first analyzes the weighted sum rate of the proposed massive access scheme, and then proposes three non-orthogonal transmit beams design algorithms. Next, Sect. 5.4 provides extensive simulation results to validate the effectiveness of the proposed schemes. Finally, Sect. 5.5 concludes the chapter.

Notations We use bold upper (lower) letters to denote matrices (column vectors), $(\cdot)^H$ to denote conjugate transpose, $E[\cdot]$ to denote expectation, $\text{var}(\cdot)$ to denote the variance, $\|\cdot\|$ to denote the L_2-norm of a vector, $\otimes$ to denote the Kronecker product, $\text{vec}(\cdot)$ to denote the vectorization of a matrix, $|\cdot|$ to denote the absolute value, and $[x]^+ = \max(x, 0)$.

5.2   System Model and Problem Formulation

We consider a single-cell cellular IoT system as shown in Fig. 5.1,[1] where a base station (BS) equipped with N_t antennas broadcasts messages to K IoT user equipments (UEs). Due to the size limitation, it is assumed that the IoT UEs have a single antenna each. Note that in the 5G cellular IoT, both the number of BS antennas N_t and the number of UEs K are very large [18]. To support massive access over limited radio spectrum, we propose a non-orthogonal beamspace multiple access framework. In what follows, we introduce the key steps of non-orthogonal beamspace multiple access framework in detail.

5.2.1   CSI Acquisition

CSI availability at the BS is the precondition of designing the massive access scheme. However, in the context of massive connections, instantaneous CSI is hardly obtained due to the limited resources. To address this issue, we propose the use of long-term channel parameters, namely statistical CSI, to design the massive access scheme. We use $\mathbf{h}_k$ to denote the N_t-dimensional channel vector from the BS to the kth UE. It is assumed that the signal sent from the BS is scattered by a cluster and then arrives at the kth UE through L propagation paths, where the lth path associated to the kth UE has a propagation distance $d_{k,l}$, a propagation attenuation

---

[1]The proposed non-orthogonal beamspace multiple access technique can be easily extended to the multiple-cell scenario.

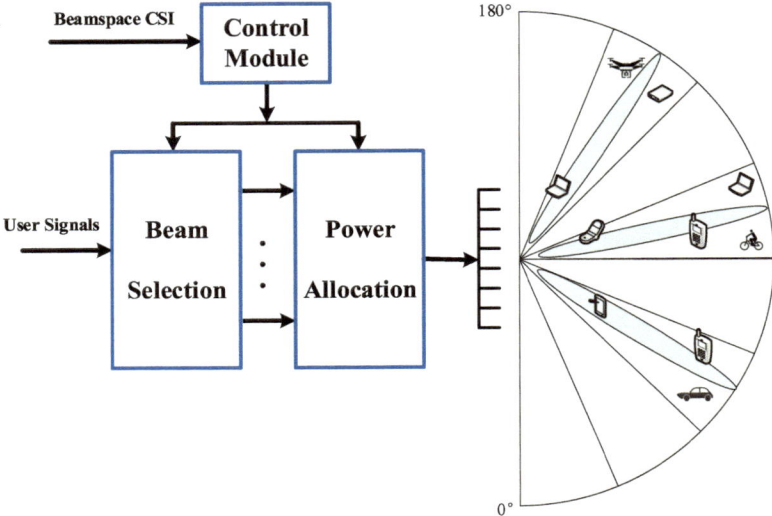

Fig. 5.1 A massive access model for the cellular IoT in the beamspace

$a_{k,l}$, and an angle of departure (AoD) $\theta_{k,l}$. Then, according to the electromagnetic wave propagation theory [14, 19], the channel vector $\mathbf{h}_k$ can be expressed as

$$\mathbf{h}_k = \sum_{l=1}^{L} a_{k,l} e^{-j2\pi d_{k,l}/\lambda_c} \mathbf{e}(\theta_{k,l}), \tag{5.1}$$

where λ_c is the carrier wavelength, and $\mathbf{e}(\theta_{k,l})$ is the transmit antenna array response vector. For a typical uniform linear array (ULA), $\mathbf{e}(\theta_{k,l})$ is given by

$$\mathbf{e}(\theta_{k,l}) = \frac{1}{\sqrt{N_t}} \left[1, \cdots, e^{-j2\pi(i-1)\varrho \sin(\theta_{k,l})}, \cdots, e^{-j2\pi(N_t-1)\varrho \sin(\theta_{k,l})} \right]^T, \forall i \in [1, N_t], \tag{5.2}$$

where ϱ is the normalized array spacing with respect to the carrier wavelength. In the considered 5G cellular IoT, since the number of BS antennas is very large, the transmit antenna array response vectors are asymptotically orthogonal, cf. [11] and [12]. Thus, with N_t antennas, the BS can separate N_t AoDs, namely $\phi_{k,i}, i = 1, \cdots, N_t$. For the ULA, it is usual to adopt uniform sampling of spatial angles, i.e., $\sin \phi_{k,i} = \frac{2i}{N_t} - 1$ with $\phi_{k,i} \in [-\frac{\pi}{2}, \frac{\pi}{2}]$. In this case, the channel vector $\mathbf{h}_k$ can be rewritten as

$$\mathbf{h}_k = \mathbf{U}_k \mathbf{\Lambda}_k^{\frac{1}{2}} \bar{\mathbf{h}}_k, \tag{5.3}$$

where $\mathbf{U}_k = [\mathbf{e}(\phi_{k,1}), \cdots, \mathbf{e}(\phi_{k,N_t})]$ denotes the transmit antenna array response matrix over N_t orthogonal base beam directions, $\boldsymbol{\Lambda}_k = \mathrm{diag}\{\eta_{k,1}, \cdots, \eta_{k,N_t}\}$ is a diagonal matrix with $\eta_{k,i}$ as the ith diagonal element being the gain in the angle $\phi_{k,i}$. In general, $\eta_{k,i}$ is determined by the sum of the propagation attenuation of multiple paths, whose angles are closest to the angle $\phi_{k,i}$. Moreover, $\bar{\mathbf{h}}_k$ is a N_t-dimensional complex Gaussian vector with zero mean and unit variance. In other words, $\mathbf{h}_k$ in (5.3) represents a linear combination of N_t base beam directions. All the combinations of these orthogonal base beams form a beamspace.

Furthermore, we consider the transmit channel correlation matrix, which can be casted as

$$\mathbf{R}_k = \mathrm{E}\left[\mathbf{h}_k \mathbf{h}_k^H\right]$$

$$= \mathbf{U}_k \boldsymbol{\Lambda}_k \mathbf{U}_k^H. \tag{5.4}$$

Given the channel correlation matrix, it is possible to obtain the gain matrix $\boldsymbol{\Lambda}_k$ over the N_t base beam directions. In fact, the channel correlation matrix remains unchanged during a relative long time and is easily obtained by statistical averaging over a large number of channel realizations. Thus, we can get the gains of the beam directions based on statistical CSI. Moreover, with the same sampling spacing of spatial angles, we have the following lemma:

Lemma 1 *If the number of BS antennas is sufficiently large, the beamspace matrix $\mathbf{U}_k$ is asymptotically identical for all the UEs, i.e.,*

$$\mathbf{U}_k \to \mathbf{U}, \forall k, \quad as \ N_t \to \infty, \tag{5.5}$$

where the ith column of $\mathbf{U}$, namely $\mathbf{u}_i$, can be expressed as $\mathbf{e}(\arcsin(\frac{2i}{N_t} - 1))$. Thus, one can use the same orthogonal bases to construct all UEs' channels in the beamspace. In summary, the BS is capable of obtaining the beamspace CSI, i.e., the base beams $\mathbf{u}_i$ and the corresponding gains $\eta_{k,i}$ of all the UEs, $\forall i \in [1, N_t], k \in [1, K]$.

5.2.2   User Clustering

User clustering can effectively reduce the computational complexity at the UEs as the SIC is only performed within a cluster, which is a critical issue for the IoT devices. Moreover, it may solve a vital problem of the BS with a large-scale antenna array, namely decreasing the number of radio-frequency (RF) chains. Specifically, if the UEs in a cluster share the same data stream, the required number of RF chains can be significantly decreased.

It is intuitive that user clustering should be performed based on the available CSI. In the proposed framework, the BS only utilizes the statistical CSI or beamspace CSI to simplify the system complexity. Now, we propose to perform user clustering in the beamspace. Generally speaking, the beamspace $[-\frac{\pi}{2}, \frac{\pi}{2}]$ is divided into N_t subspaces and a subspace corresponds to a cluster.[2] Therefore, an UE whose AoD belongs to $[-\frac{\pi}{2} + (i - 1)\frac{\pi}{N_t}, -\frac{\pi}{2} + i\frac{\pi}{N_t}]$ is grouped into the ith cluster. For the AoD information of the UEs, it can be obtained by using existing algorithms, e.g., the MUSIC algorithm [20]. Hence, we can easily partition the UEs into multiple clusters according to the AoD information. Due to the random distribution of the IoT UEs, we assume that there are M clusters[3] and N_m UEs in the mth cluster. For ease of notation, we use $\text{UE}_{m,n}$ and $\mathbf{h}_{m,n}$ to denote the nth UE in the mth cluster and the corresponding channel vector, respectively.

5.2.3 Superposition Coding

Superposition coding at the BS is a key step for achieving efficient massive access over limited radio spectrum. In general, superposition coding can be regarded as a weighted sum of the signals to be transmitted, which is mathematically given by

$$\mathbf{x} = \sum_{m=1}^{M} \sum_{n=1}^{N_m} \mathbf{w}_{m,n} x_{m,n}, \tag{5.6}$$

where $x_{m,n}$ and $\mathbf{w}_{m,n}$ is the Gaussian distributed transmit signal with unit norm and the transmit beam for the $\text{UE}_{m,n}$, respectively. In previous related works [15] and [16], the transmit beam is directly designed based on the transmit antenna array vector associated to its assigned cluster. For instance, since the $\text{UE}_{m,n}$ falls in the mth cluster, $\mathbf{w}_{m,n}$ can be simply constructed as $\mathbf{u}_i$, namely matching filter (MF) beamforming. An advantage of such a method is that the beams across the clusters are orthogonal of each other if the number of BS antennas is sufficiently large. However, due to the limited angular resolution, a beam cannot perfectly align multiple UEs in a cluster. Hence, the co-channel interference does exist even the MF beamforming is performed. To solve this problem, we propose a non-orthogonal beamspace multiple access scheme. As shown in Fig. 5.2, a transmit beam is a linear combination of multiple transmit antenna array vectors, namely base beams. In particular, the non-orthogonal beamforming multiple access scheme can significantly increase the angular resolution, and thus improves the overall

---

[2]Note that the subspace is determined by the angular resolution of the large-scale antenna array. Thus, the maximum number of clusters is N_t, but we also can use multiple subspaces to form a cluster. As a result, the number of clusters can be decreased.

[3]The system may have N_t clusters at most, but only M clusters are non-empty. Thus, we have $1 \leq M \leq N_t$.

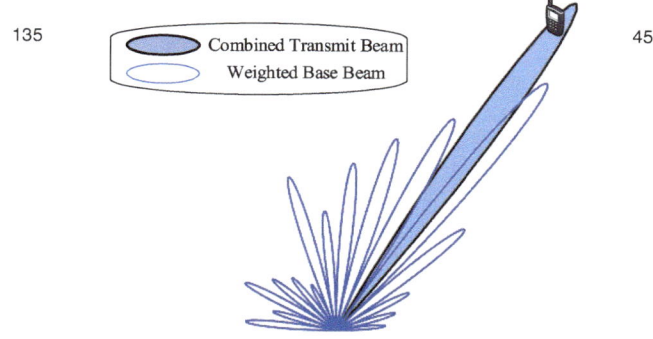

Fig. 5.2   The construction of the non-orthogonal transmit beam

performance of the cellular IoT. In this case, the transmit beam for the $\text{UE}_{m,n}$ can be expressed as

$$\mathbf{w}_{m,n} = \sum_{c \in \mathscr{B}_{m,n}} \sqrt{p_{m,n,c}} \mathbf{u}_c$$

$$= \mathbf{U} \mathbf{P}_{m,n}^{\frac{1}{2}} \mathbf{s}_{m,n}, \tag{5.7}$$

where $\mathscr{B}_{m,n}$ is the index set of selected base beams for the $\text{UE}_{m,n}$, $p_{m,n,c}$ is the transmit power on the cth base beam, and $\mathbf{P}_{m,n} = \text{diag}\{p_{m,n,1}, \cdots, p_{m,n,N_t}\}$. Moreover, $\mathbf{s}_{m,n} = [s_{m,n,1}, \cdots, s_{m,n,N_t}]^T$ is the beam selection vector. If $i \in \mathscr{B}_{m,n}$, then $s_{m,n,i} = 1$, Otherwise $s_{m,n,i} = 0$. As such, superposition coding is equivalent to the design of non-orthogonal transmit beams, namely a combination of beam selection and power allocation, which will be discussed in detail later.

5.2.4   Successive Interference Cancellation

With the superposition coded signal $\mathbf{x}$, the BS broadcasts it over the downlink channels. Without loss of generality, we consider the received signal at the $\text{UE}_{m,n}$, which is given by

$$y_{m,n} = \mathbf{h}_{m,n}^H \mathbf{x} + n_{m,n}$$

$$= \underbrace{\mathbf{h}_{m,n}^H \mathbf{w}_{m,n} x_{m,n}}_{\text{Desired signal}} + \underbrace{\sum_{i=1,i \neq n}^{N_m} \mathbf{h}_{m,n}^H \mathbf{w}_{m,i} x_{m,i}}_{\text{Intra-cluster interference}} + \underbrace{\sum_{j=1,j \neq m}^{M} \sum_{i=1}^{N_j} \mathbf{h}_{m,n}^H \mathbf{w}_{j,i} x_{j,i}}_{\text{Inter-cluster interference}} + \underbrace{n_{m,n}}_{\text{AWGN}} ,$$

$$(5.8)$$

where $n_{m,n}$ is additive white Gaussian noise (AWGN) with unit variance. Note that the first term at the right side of Eq. (5.8) is the desired signal, the second one is the intra-cluster interference, and the third one is the inter-cluster interference. In order to further improve the quality of the received signal, the $UE_{m,n}$ carries out SIC within the associated cluster. Without loss of generality, it is assumed that the desired channel gains in the mth cluster has a descending order as follows

$$g_{m,1} \geq \cdots g_{m,n} \geq g_{m,N_m}, \qquad (5.9)$$

where $g_{m,n}$ is defined as

$$g_{m,n} = |\mathbf{h}_{m,n}^H \mathbf{w}_{m,n}|^2$$

$$= \left| \mathbf{h}_{m,n}^H \Lambda_{m,n}^{\frac{1}{2}} \mathbf{U}^H \mathbf{U} \mathbf{P}_{m,n}^{\frac{1}{2}} \mathbf{s}_{m,n} \right|^2$$

$$= |\bar{\mathbf{h}}_{m,n}^H \Lambda_{m,n}^{\frac{1}{2}} \mathbf{P}_{m,n}^{\frac{1}{2}} \mathbf{s}_{m,n}|^2, \qquad (5.10)$$

with $\bar{\mathbf{h}}_{m,n}$ being the unknown small-scale channel fading vector, and $\Lambda_{m,n} = \text{diag}\{\eta_{m,n,1}, \cdots, \eta_{m,n,N_t}\}$ being the gain matrix of the orthogonal base beams for the $UE_{m,n}$. In practice, each UE knows its channel gain through channel estimation for coherent detection [21]. Thus, the UEs can feed back the channel gains to the BS via the uplink and the BS determines the order of the channel gains in each cluster which is informed to the UEs through the downlink. Based on the order of the channel gains, the $UE_{m,n}$ decodes the interfering signals and subtracts the corresponding interference from the N_mth to $(n + 1)$th UE in sequence, and finally demodulates its desired signal $x_{m,n}$. Hence, the received signal after SIC at the $UE_{m,n}$ can be expressed as

$$y'_{m,n} = \mathbf{h}_{m,n}^H \mathbf{w}_{m,n} x_{m,n} + \sum_{i=1}^{n-1} \mathbf{h}_{m,n}^H \mathbf{w}_{m,i} x_{m,i} + \sum_{j=1,j \neq m}^{M} \sum_{i=1}^{N_j} \mathbf{h}_{m,n}^H \mathbf{w}_{j,i} x_{j,i} + n_{m,n},$$

$$(5.11)$$

and the corresponding signal-to-interference-plus-noise ratio (SINR) can be computed as

$$
\gamma_{m,n} = \frac{|\mathbf{h}_{m,n}^{H}\mathbf{w}_{m,n}|^2}{\sum\limits_{i=1}^{n-1}|\mathbf{h}_{m,n}^{H}\mathbf{w}_{m,i}|^2 + \sum\limits_{j=1,j\neq m}^{M}\sum\limits_{i=1}^{N_j}|\mathbf{h}_{m,n}^{H}\mathbf{w}_{j,i}|^2 + 1}
$$

$$
= \frac{|\bar{\mathbf{h}}_{m,n}^{H}\Lambda_{m,n}^{\frac{1}{2}}\mathbf{P}_{m,n}^{\frac{1}{2}}\mathbf{s}_{m,n}|^2}{\sum\limits_{i=1}^{n-1}|\bar{\mathbf{h}}_{m,n}^{H}\Lambda_{m,n}^{\frac{1}{2}}\mathbf{P}_{m,i}^{\frac{1}{2}}\mathbf{s}_{m,i}|^2 + \sum\limits_{j=1,j\neq m}^{M}\sum\limits_{i=1}^{N_j}|\bar{\mathbf{h}}_{m,n}^{H}\Lambda_{m,n}^{\frac{1}{2}}\mathbf{P}_{j,i}^{\frac{1}{2}}\mathbf{s}_{j,i}|^2 + 1}.
$$

$$(5.12)$$

Thus, the achievable rate of the $UE_{m,n}$ is given by

$$
r_{m,n} = \log_2(1 + \gamma_{m,n})
$$

$$
= \log_2\left(\frac{\sum\limits_{j=1,j\neq m}^{M}\sum\limits_{i=1}^{N_j}|\bar{\mathbf{h}}_{m,n}^{H}\Lambda_{m,n}^{\frac{1}{2}}\mathbf{P}_{j,i}^{\frac{1}{2}}\mathbf{s}_{j,i}|^2 + \sum\limits_{i=1}^{n}|\bar{\mathbf{h}}_{m,n}^{H}\Lambda_{m,n}^{\frac{1}{2}}\mathbf{P}_{m,i}^{\frac{1}{2}}\mathbf{s}_{m,i}|^2 + 1}{\sum\limits_{j=1,j\neq m}^{M}\sum\limits_{i=1}^{N_j}|\bar{\mathbf{h}}_{m,n}^{H}\Lambda_{m,n}^{\frac{1}{2}}\mathbf{P}_{j,i}^{\frac{1}{2}}\mathbf{s}_{j,i}|^2 + \sum\limits_{i=1}^{n-1}|\bar{\mathbf{h}}_{m,n}^{H}\Lambda_{m,n}^{\frac{1}{2}}\mathbf{P}_{m,i}^{\frac{1}{2}}\mathbf{s}_{m,i}|^2 + 1}\right).
$$

$$(5.13)$$

From (5.13), the achievable rate is mainly determined by the beam selection and the corresponding power allocation, namely the design of non-orthogonal transmit beams. To improve the system performance, we design the non-orthogonal beamforming schemes according to the characteristics of the beamspace massive access system.

5.3 Performance Analysis and Optimization of Non-orthogonal Beamspace Multiple Access

In this section, we aim to design the non-orthogonal beamspace multiple access schemes from the perspective of maximizing the weighted sum of the ergodic rates of the cellular IoT. To facilitate the analysis, we first analyze the weighted sum of the ergodic rates.

5.3.1 Performance Analysis

In general, the weighted sum of the ergodic rate can be expressed as

$$R_{\text{sum}} = \sum_{m=1}^{M} \sum_{n=1}^{N_m} \alpha_{m,n} \mathbb{E}[r_{m,n}], \tag{5.14}$$

where $\alpha_{m,n} > 0$ denote the priority of the $UE_{m,n}$. As seen in (5.13), $r_{m,n}$ is a complicated function of the random variable $\|\bar{\mathbf{h}}_{m,n}\|^2$, thus it is difficult to obtain a closed-form expression for the weighted sum rate R_{sum}. As an alternative, we derive an upper bound on the weighted sum of the ergodic rates, which is summarized in the following theorem:

Theorem 1 *Based on the proposed non-orthogonal beamspace multiple access framework, the weighted sum of the ergodic rates of all UEs is upper bounded by*

$$R_{\text{ub}} = \sum_{m=1}^{M} \sum_{n=1}^{N_m} \sum_{c=1}^{N_t} \alpha_{m,n} \left(\log_2 \left(\left(\sum_{j=1, j \neq m}^{M} \sum_{i=1}^{N_j} s_{j,i,c} p_{j,i,c} + \sum_{i=1}^{n} s_{m,i,c} p_{m,i,c} \right) \eta_{m,n,c} + 1 \right) \right.$$

$$\left. - \log_2 \left(\left(\sum_{j=1, j \neq m}^{M} \sum_{i=1}^{N_j} s_{j,i,c} p_{j,i,c} + \sum_{i=1}^{n-1} s_{m,i,c} p_{m,i,c} \right) \eta_{m,n,c} + 1 \right) \right). \tag{5.15}$$

Proof Please refer to "Appendix The Proof of Theorem 1".

Note that the upper bound R_{ub} can be rewritten as

$$R_{\text{ub}} = \sum_{m=1}^{M} \sum_{n=1}^{N_m} \sum_{c=1}^{N_t} \alpha_{m,n} \log_2 \left(1 + \frac{s_{m,n,c} p_{m,n,c} \eta_{m,n,c}}{\sum\limits_{j=1, j \neq m}^{M} \sum\limits_{i=1}^{N_j} s_{j,i,c} p_{j,i,c} \eta_{m,n,c} + \sum\limits_{i=1}^{n-1} s_{m,i,c} p_{m,i,c} \eta_{m,n,c} + 1} \right). \tag{5.16}$$

It is seen that the upper bound R_{ub} is equivalent to a sum rate on N_t orthogonal resource blocks. As such, there is no interference among these resource blocks. Furthermore, according to the upper bound, we can obtain the following proposition:

Proposition 1 *In the scenario of massive connections, the weighted sum of the ergodic rates will be saturated.*

Proof Let $p_{j,i,c} = v_{j,i,c} P_{\text{tot}}$, where P_{tot} is the total transmit power of the BS, and $v_{j,i,c}$ is the power allocation factor for the $UE_{j,i}$ over the cth base beam. In the scenario of massive connections, if the transmit power is high enough, the cellular IoT usually interference-limited. In other words, the noise is negligible compared to the interference. Then, the upper bound is reduced to

$$R_{\mathrm{ub}} \approx \sum_{m=1}^{M}\sum_{n=1}^{N_m}\sum_{c=1}^{N_t} \alpha_{m,n} \log_2 \left(1 + \frac{s_{m,n,c}v_{m,n,c}P_{\mathrm{tot}}\eta_{m,n,c}}{\sum_{j=1,j\neq m}^{M}\sum_{i=1}^{N_j} s_{j,i,c}v_{j,i,c}P_{\mathrm{tot}}\eta_{m,n,c} + \sum_{i=1}^{n-1} s_{m,i,c}v_{m,i,c}P_{\mathrm{tot}}\eta_{m,n,c}} \right)$$

$$= \sum_{m=1}^{M}\sum_{n=1}^{N_m}\sum_{c=1}^{N_t} \alpha_{m,n} \log_2 \left(1 + \frac{s_{m,n,c}v_{m,n,c}}{\sum_{j=1,j\neq m}^{M}\sum_{i=1}^{N_j} s_{j,i,c}v_{j,i,c} + \sum_{i=1}^{n-1} s_{m,i,c}v_{m,i,c}} \right). \qquad (5.17)$$

It is seen in (5.17) that the upper bound is independent of the transmit power, and thus it will be saturated. In addition, the saturated upper bound also has nothing to do with the channel gains in the beamspace. Thus, the UEs with the same order but in different cluster may asymptotically achieve the same performance.

5.3.2   Performance Optimization

From Theorem 1, the overall performance of the cellular IoT depends on the design of non-orthogonal beams in the beamspace, namely beam selection and power allocation. In this section, we jointly optimize beam selection and power allocation from the perspective of maximizing the upper bound on the weighted sum of the ergodic rates, which can be formulated as the following problem:

$$J_1 : \max_{\mathbf{S},\mathbf{P}} R_{\mathrm{ub}}$$

$$\text{s.t. C1} : \sum_{m=1}^{M}\sum_{n=1}^{N_m}\sum_{c=1}^{N_t} s_{m,n,c}p_{m,n,c} \leq P_{\max},$$

where $P_{\max}$ is the maximum transmit power at the BS, $\mathbf{S} = \{s_{1,1,1}, \cdots, s_{M,N_M,N_t}\}$ and $\mathbf{P} = \{p_{1,1,1}, \ldots, p_{M,N_M,N_t}\}$ are the beam selection and power allocation matrices, respectively. Since $s_{m,n,c}, \forall m, n, c$ should be in $\{0, 1\}$, J_1 is a mixed integer programming problem. Thus, it is difficult to obtain the optimal solutions.

Generally speaking, beam selection forms the set of the orthogonal base beams for serving the UEs, and power allocation constructs the transmit beams based on the selected base beams. Thus, the number of the available orthogonal base beams determines the size of the subspace, namely the angular resolution. Inspired by this, we provide three suboptimal design algorithms from the viewpoint of the angular resolution in the beamspace.

5.3.2.1 Full-Space Multiple-Beam Design

First, we consider the case that the BS can select all orthogonal base beams for each UE, namely $s_{m,n,c} = 1, \forall m, n, c$. As seen in Fig. 5.3, the beam for each UE can be constructed in the whole beamspace, and the transmit beams are in general non-orthogonal. In this case, the beam design is reduced to the problem of power allocation, namely determining the weighted coefficients for each UE's orthogonal base beams, which can be expressed as the following problem:

$$J_2 : \max_{\mathbf{P}} \sum_{m=1}^{M} \sum_{n=1}^{N_m} \sum_{c=1}^{N_t} \alpha_{m,n} \log_2 \left(1 + \frac{p_{m,n,c}\eta_{m,n,c}}{\sum_{j=1,j\neq m}^{M} \sum_{i=1}^{N_j} p_{j,i,c}\eta_{m,n,c} + \sum_{i=1}^{n-1} p_{m,i,c}\eta_{m,n,c} + 1} \right)$$

$$\text{s.t. } C2 : \sum_{m=1}^{M} \sum_{n=1}^{N_m} \sum_{c=1}^{N_t} p_{m,n,c} \leq P_{\max}.$$

Unfortunately, the objective function of J_2 is still non-concave with respect to the transmit powers, and it is difficult to design optimal power allocation. Checking the objective function of J_2, it is found that it can be considered as the sum rate of KN_t independent UEs, where the transmit power is $p_{m,n,c}$ and the channel gain is $\eta_{m,n,c}$ for the equivalent $UE_{m,n,c}$. Therefore, we have the following equivalent input-output relation to the $UE_{m,n,c}$:

$$y_{m,n,c} = \sqrt{p_{m,n,c}\eta_{m,n,c}}x_{m,n,c} + \sqrt{\sum_{j=1,j\neq m}^{M} \sum_{i=1}^{N_j} p_{j,i,c}\eta_{m,n,c} + \sum_{i=1}^{n-1} p_{m,i,c}\eta_{m,n,c}x'_{m,n,c}} + n_{m,n,c},$$

$$(5.18)$$

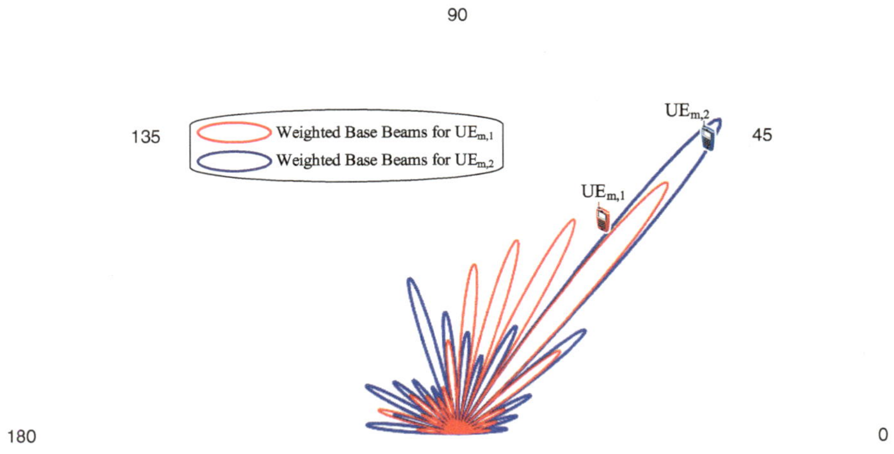

Fig. 5.3 The multiple-beam design in the whole beamspace

where $x_{m,n,c}$ and $x'_{m,n,c}$ are the normalized Gaussian distributed desired signal and the interfering signal, respectively, and $n_{m,n,c}$ is the AWGN with unit norm. Then, the equivalent rate for the $UE_{m,n,c}$ is given by

$$R_{m,n,c} = \log_2 \left(1 + \underbrace{\frac{p_{m,n,c}\eta_{m,n,c}}{\sum\limits_{j=1,j\neq m}^{M}\sum\limits_{i=1}^{N_j} p_{j,i,c}\eta_{m,n,c} + \sum\limits_{i=1}^{n-1} p_{m,i,c}\eta_{m,n,c} + 1}}_{\gamma_{m,n,c}} \right), \quad (5.19)$$

where $\gamma_{m,n,c}$ is the equivalent SINR for the equivalent $UE_{m,n,c}$. It is well known that the minimum mean squared error (MSE) and the SINR for an arbitrary received signal have the following relation [22, 23]:

Lemma 2 *The minimum MSE $e_{m,n,c}$ and the SINR $\gamma_{m,n,c}$ of a received signal satisfy*

$$e_{m,n,c}^{-1} = 1 + \gamma_{m,n,c}. \quad (5.20)$$

In other words, maximizing the rate is equivalent to minimizing the MSE. According to the input-output relation in (5.18), the MSE for the equivalent $UE_{m,n,c}$ can be easily expressed as

$$
\begin{aligned}
MSE_{m,n,c} &= E[(v_{m,n,c}y_{m,n,c} - x_{m,n,c})(v_{m,n,c}y_{m,n,c} - x_{m,n,c})^H] \\
&= v_{m,n,c}\left(\eta_{m,n,c}\left(\sum_{j=1,j\neq m}^{M}\sum_{i=1}^{N_j} p_{j,i,c} + \sum_{i=1}^{n} p_{m,i,c}\right) + 1\right)v_{m,n,c}^H \\
&\quad - \sqrt{\eta_{m,n,c}p_{m,n,c}}\left(v_{m,n,c} + v_{m,n,c}^H\right) + 1 \\
&= \left(v_{m,n,c} - \sqrt{\eta_{m,n,c}p_{m,n,c}}\Phi_{m,n,c}^{-1}\right)\Phi_{m,n,c}\left(v_{m,n,c} - \sqrt{\eta_{m,n,c}p_{m,n,c}}\Phi_{m,n,c}^{-1}\right)^H \\
&\quad - \eta_{m,n,c}p_{m,n,c}\Phi_{m,n,c}^{-1} + 1,
\end{aligned}
\quad (5.21)
$$

where $v_{m,n,c}$ denotes the receiver, and $\Phi_{m,n,c} = \eta_{m,n,c}\left(\sum\limits_{j=1,j\neq m}^{M}\sum\limits_{i=1}^{N_j} p_{j,i,c} + \sum\limits_{i=1}^{n} p_{m,i,c}\right) + 1$ is the power of the received signal of the equivalent $UE_{m,r,c}$. It is clear that $MSE_{m,n,c}$ is minimized if and only if $v_{m,n,c} = \sqrt{\eta_{m,n,c}p_{m,n,c}}\Phi_{m,n,c}^{-1}$, namely adopting the MMSE receiver.

Thus, according to Lemma 2, the objective function of J_2 can be transformed as

$$\min_{\mathbf{P}} \sum_{m=1}^{M} \sum_{n=1}^{N_m} \sum_{c=1}^{N_t} \alpha_{m,n} (\log_2(e_{m,n,c})). \tag{5.22}$$

However, the objective function in (5.22) is still not convex. Note that (5.22) aims to minimize a function of minimum MSE, which is equivalent to minimizing a function of MSE for a given MMSE receiver. In other words, the optimization objective in (5.22) can be transformed as

$$\min_{\mathbf{P},\mathbf{v}} \sum_{m=1}^{M} \sum_{n=1}^{N_m} \sum_{c=1}^{N_t} \alpha_{m,n} (\log_2(\text{MSE}_{m,n,c})), \tag{5.23}$$

where $\mathbf{v} = \{v_{1,1,1}, \cdots, v_{M,N_M,N_t}\}$ is the collection of the receivers. The sum of logarithmic functions hinders us to further solve this problem. Similar to [22] and [23], we can replace the logarithmic function with the following term

$$\min_{\mathbf{P},\mathbf{v},\boldsymbol{\beta}} \sum_{m=1}^{M} \sum_{n=1}^{N_m} \sum_{c=1}^{N_t} \alpha_{m,n} (\beta_{m,n,c} \text{MSE}_{m,n,c} - \log_2(\beta_{m,n,c})), \tag{5.24}$$

where $\boldsymbol{\beta} = \{\beta_{1,1,1}, \cdots, \beta_{M,N_M,N_t}\}$ is the collection of auxiliary variables. Note that only when $\beta_{m,n,c} = \text{MSE}_{m,n,c}^{-1}$, the objective function in (5.24) can achieve its minimum value. Under such a condition, the optimization objectives (5.23) and (5.24) are equivalent.

According to the definition of $\text{MSE}_{m,n,c}$ in (5.21), it is known that (5.24) is not a jointly convex function of $\mathbf{P}$, $\mathbf{v}$, and $\boldsymbol{\beta}$, but it is a convex function with respect to each optimization variable. Thus, we can adopt the sequential iteration optimization method to solve the problem. Specifically, we can optimize one variable by fixing the others, and the variables are iteratively optimized until they approach a stationary point. First, for the variable $\mathbf{P}$, by combining the objective function (5.24) and the constraint condition C2, we get the Lagrange function as

$$\mathscr{L}_1(\mathbf{P}) = \sum_{m=1}^{M} \sum_{n=1}^{N_m} \sum_{c=1}^{N_t} \alpha_{m,n} (\beta_{m,n,c} \text{MSE}_{m,n,c} - \log_2(\beta_{m,n,c})) + \mu \left(\sum_{m=1}^{M} \sum_{n=1}^{N_m} \sum_{c=1}^{N_t} P_{m,n,c} - P_{\max} \right), \tag{5.25}$$

where $\mu \geq 0$ is the Lagrange multiplier of C2. By leveraging the KKT conditions, we obtain

$$p_{m,n,c} = \left(\frac{\alpha_{m,n} \beta_{m,n,c} v_{m,n,c} \sqrt{\eta_{m,n,c}}}{\sum_{j=1,j \neq m}^{M} \sum_{i=1}^{N_m} \alpha_{j,i} \beta_{j,i,c} v_{j,i,c}^2 \eta_{j,i,c} + \sum_{i=1}^{n} \alpha_{m,i} \beta_{m,i,c} v_{m,i,c}^2 \eta_{m,i,c} + \mu} \right)^2. \tag{5.26}$$

Then, the intermediate variables $\mathbf{v}$ and β capturing the performance of the UEs in the previous iteration can be obtained according to their definitions. Moreover, the Lagrange multiplier μ can be updated by the iterative gradient method. In the $(t + 1)$th iteration, the μ can be updated as

$$\mu(t+1) = \left[\mu(t) + \Delta_\mu \left(\sum_{m=1}^{M} \sum_{n=1}^{N_m} \sum_{c=1}^{N_t} p_{m,n,c} - P_{\max} \right) \right]^+, \tag{5.27}$$

where $\Delta_\mu > 0$ is an iteration step size. In summary, the full-space multiple-beam design algorithm can be described as

---

Algorithm 1 Full-space multiple-beam design

Input:	$P_{\max}, \alpha_{m,n}, \eta_{m,n,c}, \forall m, n, c$;
Step 1:	Initialize the parameters by setting $p_{m,n,c} = \frac{P_{\max}}{K N_t}, \forall m, n, c$;
Step 2:	$\mu = 1, v_{m,n,c} = \sqrt{\eta_{m,n,c} p_{m,n,c}} \Phi_{m,n,c}^{-1}, \beta_{m,n,c} = \mathrm{MSE}_{m,n,c}^{-1}, \forall m, n, c$;
Step 3:	Update $p_{m,n,c}$ according to (5.26), $\forall m, n, c$;
Step 4:	Update μ by the gradient method in (5.27). If μ is not converged, go to Step 3;
Step 5:	If $\mathbf{P}$ is not converged, go to Step 2;
Output:	$\mathbf{P}$.

In this algorithm, we construct the transmit beam in the whole beamspace, and thus there are $K N_t$ optimization variables for $\mathbf{P}$ in total. In fact, since the UEs are partitioned into clusters according to the AoD information, the transmit beam for an arbitrary UE is mainly determined by a finite number of base beams with high correlation in the beamspace. Thus, the required number of optimization variables for $\mathbf{P}$ can be effectively reduced.

5.3.2.2 Partial-Space Multiple-Beam Design

In this section, we design the transmit beam for each UE with a few base beams, namely in the partial beamspace, cf. Fig. 5.4. Note that in the scenario of massive access for the cellular IoT, the system performance is mainly limited by the co-channel interference. In order to effectively reduce the co-channel interference, we enforce the subspaces between the clusters orthogonal to each other. In specific, the N_t base beams are allocated to the M clusters, and a base beam can only serve the UEs in a cluster, namely $\sum_{m=1}^{M} s_{m,n,c} = 1$. As a result, we have

$$\sum_{j=1, j \neq m}^{M} \sum_{i=1}^{N_j} s_{j,i,c} p_{j,i,c} \eta_{m,n,c} = 0, \forall j \neq m.$$ In this case, the beam design can be formulated as the following problem

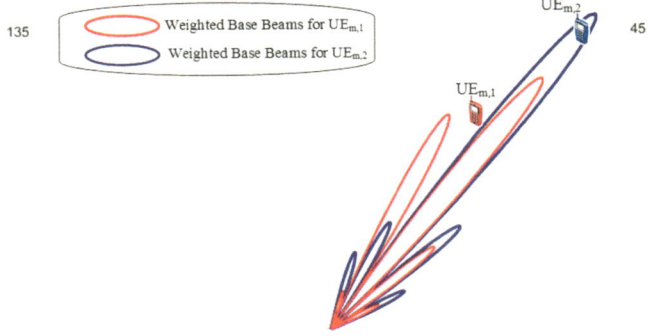

Fig. 5.4 The multiple-beam design in the partial beamspace

$$J_3 : \max_{\mathbf{S},\mathbf{P}} \sum_{m=1}^{M} \sum_{n=1}^{N_m} \sum_{c=1}^{N_t} \alpha_{m,n} \log_2 \left(1 + \frac{s_{m,n,c} p_{m,n,c} \eta_{m,n,c}}{\sum_{i=1}^{n-1} s_{m,i,c} p_{m,i,c} \eta_{m,n,c} + 1} \right)$$

$$\text{s.t. C1, C3} : \sum_{m=1}^{M} s_{m,n,c} = 1, \forall n, c.$$

As mentioned above, due to the constraint of $s_{m,n,c} \in \{0, 1\}$, J_3 is also a mixed integer programming problem. To solve this problem, we partition it into two subproblems, one for beam selection, and the other for power allocation.

We first address the problem of beam selection. It is intuitive that the optimal beam selection can be realized by the exhaustive searching. However, since there might be a massive number of clusters and base beams, the computational complexity of the exhaustive searching is prohibitive. Hence, we propose a low-complexity beam selection method according to the characteristics of massive access in the beamspace. Checking the objective function of J_3, it is found that since a base beam is only distributed to one cluster exclusively, there is no inter-cluster interference. In other words, the clusters are independent of each other over an arbitrary base beam. Meanwhile, there is no interrelation among the base beams. Thus, we can allocate the base beams one by one. For a certain base beam, in order to maximize the weighted sum rate, it is better to allocate it to the cluster with the maximum weighted sum rate over such a base beam. Mathematically, the beam selection method can be expressed as

$$m^\star = \arg \max_{m=1,\cdots,M} \sum_{n=1}^{N_m} \alpha_{m,n} \log_2 \left(1 + \frac{p_{m,n,c} \eta_{m,n,c}}{\sum_{i=1}^{n-1} p_{m,i,c} \eta_{m,n,c} + 1} \right). \qquad (5.28)$$

Hence, the cth base beam is allocated to the $m^\star$th cluster.

Then, given the beam selection result, we allocate the power for constructing the transmit beam of each UE. Fortunately, based on the selected base beams, power allocation is similar to the optimization problem J_2. Therefore, with the same method, if the cth is allocated to the mth cluster, the transmit power $p_{m,n,c}$ is given by

$$p_{m,n,c} = \left(\frac{\alpha_{m,n,c} \beta'_{m,n,c} v'_{m,n,c} \sqrt{\eta_{m,n,c}}}{\sum_{i=1}^{n} \alpha_{m,i,c} \beta'_{m,i,c} (v'_{m,i,c})^2 \eta_{m,i,c} + \mu'} \right)^2, \qquad (5.29)$$

where

$$v'_{m,n,c} = \sqrt{\eta_{m,n,c} p_{m,n,c}} \left(\eta_{m,n,c} \sum_{i=1}^{n} p_{m,i,c} + 1 \right)^{-1}, \qquad (5.30)$$

$$\beta'_{m,n,c} = \left(v'_{m,n,c} \left(\eta_{m,n,c} \sum_{i=1}^{n} p_{m,i,c} + 1 \right) (v'_{m,n,c})^H - \sqrt{\eta_{m,n,c} p_{m,n,c}} \left(v'_{m,n,c} + (v'_{m,n,c})^H \right) + 1 \right)^{-1}, \qquad (5.31)$$

and the Lagrange multiplier μ' can be updated as

$$\mu'(t+1) = \left[\mu'(t) + \Delta_{\mu'} \left(\sum_{m=1}^{M} \sum_{n=1}^{N_m} \sum_{c=1}^{N_t} s_{m,n,c} p_{m,n,c} - P_{\max} \right) \right]^+, \qquad (5.32)$$

where $\Delta_{\mu'} > 0$ is an iteration step size. Thus, the partial-space multiple-beam design algorithm can be summarized as

Algorithm 2 Partial-space multiple-beam design

Input:	$P_{\max}, \alpha_{m,n}, \eta_{m,n,c}, \forall m, n, c$;
Step 1:	Initialize the parameters by setting $s_{m,n,c} = 0$, $p_{m,n,c} = \frac{P_{\max}}{N_m N_t}$, $\forall m, n, c, k = 1$;
Step 2:	Choose $m^\star$ and let $s_{m^\star,n,k} = 1$ according to (5.28),$\forall n$, and let $p_{j,n,k} = 0, \forall j \neq m^\star, n$;
Step 3:	If $k < N_t$, then let $k = k + 1$, and go to Step 2;
Step 4:	$\mu' = 1$, $v'_{m,n,c}$ and $\beta'_{m,n,c}$ are updated according to (5.30) and (5.31), respectively, $\forall m, n, c$;
Step 5:	Update $p_{m,n,c}$ according to (5.29), $\forall m, n, c$;
Step 6:	Update μ' by the gradient method in (5.32). If μ' is not converged, go to Step 5;
Step 7:	If $\mathbf{P}$ is not converged, go to Step 4;
Output:	$\mathbf{S}, \mathbf{P}$.

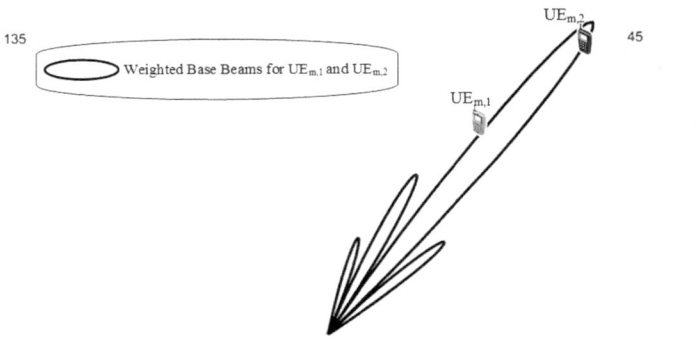

Fig. 5.5 The single-beam design in the partial beamspace

In this algorithm, we construct a transmit beam for each UE. Thus, we need to have K transmit beams in total. In the scenario of massive access for the cellular IoT, the required number of transmit beams might be very large, resulting in a high computational complexity. In fact, since we perform user clustering according to the AoD information, the UEs in a cluster may have a small difference in the AoD. Especially in the case of a large-scale antenna array at the BS, the AoDs of the UEs in a cluster are nearly the same. In this context, we can construct the same transmit beam for a cluster, and hence the required number of transmit beams and the corresponding computational complexity can be reduced significantly.

5.3.2.3   Partial-Space Single-Beam Design

As mentioned above, considering the high correlation among the UEs in a cluster, we can design only one transmit beam for a cluster. Meanwhile, to mitigate the inter-cluster interference, the subspaces, namely the base beams, for designing the transmit beams are orthogonal of each other. In general, as shown in Fig. 5.5, the total N_t base beams are divided into M sets, and each set of base beams is used to construct a transmit beam for a specific cluster.

It is assumed that the transmit beam for the mth cluster is $\mathbf{w}_m$, then it can be constructed as

$$\mathbf{w}_m = \sum_{c \in \mathscr{B}_m} \sqrt{p_c}\mathbf{u}_c, \tag{5.33}$$

where $\mathscr{B}_m$ is the index collection of the selected base beams for the mth cluster, and p_c is the total transmit power over the cth base beam. In order to guarantee that $\mathbf{w}_m$ and $\mathbf{w}_{m,n}$, $\forall n$ are aligned, the transmit power of the $\text{UE}_{m,n}$ over the cth base beam $p_{m,n,c}$ should satisfy the following condition

$$p_{m,n,c} = \iota_{m,n}p_c, \tag{5.34}$$

where $0 \le \iota_{m,n} \le 1$ is the power allocation factor of the $\text{UE}_{m,n}$ with the constraint $\sum_{n=1}^{N_m} \iota_{m,n} = 1$. Then, the relation between $\mathbf{w}_m$ and $\mathbf{w}_{m,n}$ can be expressed as

$$\mathbf{w}_{m,n} = \sqrt{\iota_{m,n}}\mathbf{w}_m = \sqrt{\iota_{m,n}} \sum_{c \in \mathscr{B}_m} \sqrt{p_c}\mathbf{u}_c = \sum_{c \in \mathscr{B}_m} \sqrt{p_{m,n,c}}\mathbf{u}_c, \tag{5.35}$$

which satisfies the power equality, and $\mathbf{w}_{m,n}$ is aligned with $\mathbf{w}_m$.

Thus, the partial-space single-beam design can be formulated as an optimization problem as below

$$J_4 : \max_{\mathbf{S},\mathbf{p}_c,\iota} \sum_{m=1}^{M}\sum_{n=1}^{N_m}\sum_{c=1}^{N_t} \alpha_{m,n} \log_2 \left(1 + \frac{s_{m,n,c}p_c\iota_{m,n}\eta_{m,n,c}}{s_{m,n,c}p_c\eta_{m,n,c}\sum_{i=1}^{n-1} \iota_{m,i} + 1} \right)$$

$$\text{s.t. C3, C4} : \sum_{c=1}^{N_t} p_c \le P_{\max},$$

$$\text{s.t. C5} : \sum_{n=1}^{N_m} \iota_{m,n} = 1, \forall m,$$

where $\mathbf{p}_c = \{p_1, \ldots, p_c\}$ and $\iota = \{\iota_{1,1}, \cdots, \iota_{M,N_M}\}$. Similarly, J_4 is a mixed integer programming problem, and we solve it through beam selection and power allocation separately. For beam selection, we can adopt the same method as (5.28) in the last algorithm.

Then, we design the power allocation method as before. First, given the beam selection result, the Lagrange function of J_4 can be written as

$$\mathscr{L}_2(\mathbf{p}_c, \boldsymbol{\iota}) = \sum_{m=1}^{M} \sum_{n=1}^{N_m} \sum_{c=1}^{N_t} \alpha_{m,n} (\beta''_{m,n,c} \mathrm{MSE}''_{m,n,c} - \log_2(\beta''_{m,n,c}))$$

$$+\mu'' \left(\sum_{c=1}^{N_t} p_c - P_{\max} \right) + \sum_{m=1}^{M} \omega_m \left(\sum_{n=1}^{N_m} \iota_{m,n} - 1 \right), \quad (5.36)$$

where

$$\mathrm{MSE}''_{m,n,c} = v''_{m,n,c} \left(\eta_{m,n,c} p_c \sum_{i=1}^{n} \iota_{m,i} + 1 \right) (v''_{m,n,c})^H - \sqrt{\eta_{m,n,c} p_c \iota_{m,n}} \left(v''_{m,n,c} + (v''_{m,n,c})^H \right) + 1,$$
$$\tag{5.37}$$

$$v''_{m,n,c} = \sqrt{\eta_{m,n,c} p_c \iota_{m,n}} \left(\eta_{m,n,c} p_c \sum_{i=1}^{n} \iota_{m,i} + 1 \right)^{-1}, \quad (5.38)$$

and $\beta''_{m,n,c} = (\mathrm{MSE}''_{m,n,c})^{-1}$. Moreover, $\mu'' \geq 0$ and $\omega_m \geq 0$ are the Lagrange multipliers of C4 and C5, respectively. Then, by using the KKT conditions, we have

$$p_c = \left(\frac{\sum_{i=1}^{N_{m^\star}} \alpha_{m^\star,i} \beta''_{m^\star,i,c} v''_{m^\star,i,c} \sqrt{\eta_{m^\star,i,c} \iota_{m^\star,i}}}{\mu'' + \sum_{i=1}^{N_{m^\star}} \alpha_{m^\star,i} \beta''_{m^\star,i,c} (v''_{m^\star,i,c})^2 \eta_{m^\star,i,c} \sum_{q=1}^{i} \iota_{m^\star,q}} \right)^2. \quad (5.39)$$

and

$$\iota_{m,n} = \left(\frac{\sum_{c=1}^{N_t} s_{m,n,s} \alpha_{m,n} \beta''_{m,n,c} v''_{m,n,c} \sqrt{\eta_{m,n,c} p_c}}{\sum_{c=1}^{N_t} \sum_{i=n}^{N_m} s_{m,n,s} \alpha_{m,i} \beta''_{m,i,c} (v''_{m,i,c})^2 \eta_{m,i,c} p_c + \omega_n} \right)^2, \quad (5.40)$$

where $m^\star$ is the index of the cluster which uses the cth base beam according to (5.28). Moreover, μ'' and ω_m can be updated by the following gradient methods

$$\mu''(t+1) = \left[\mu''(t) + \Delta_{\mu''} \left(\sum_{c=1}^{N_t} p_c - P_{\max} \right) \right]^+, \quad (5.41)$$

and

$$\omega_m(t+1) = \left[\omega_m(t) + \Delta_{\omega_m} \left(\sum_{n=1}^{N_m} \iota_{m,n} - 1 \right) \right]^+, \forall m, \tag{5.42}$$

where $\Delta_{\mu''} > 0$ and $\Delta_{\omega_m} > 0$ are iteration step sizes. Thus, the partial-space single beam design algorithm for massive access can be summarized as

---

Algorithm 3 Partial-space single-beam design

Input: $P_{\max}, \alpha_{m,n}, \eta_{m,n,c}, \forall m, n, c$;

Step 1: Initialize the parameters by setting $s_{m,n,c} = 0$, $p_c = \frac{P_{\max}}{N_t}$, $\iota_{m,n} = \frac{1}{N_m}, \forall m, n. c$, and $k = 1$;

Step 2: Choose $m^\star$ and let $s_{m^\star,n,k} = 1$ according to (5.28),$\forall n$;

Step 3: If $k < N_t$, then let $k = k + 1$, and go to Step 2;

Step 4: $\mu'' = 1$, $\omega_m = 1$, $v_{m,n,c}'' = \sqrt{\eta_{m,n,c} p_c \iota_{m,n}} \left(\eta_{m,n,c} p_c \sum_{i=1}^{n} \iota_{m,i} + 1 \right)^{-1}$, $\beta_{m,n,c}'' = $ $(\text{MSE}_{m,n,c}'')^{-1}, \forall m, n, c$;

Step 5: Update p_c according to (5.39), $\forall c$;

Step 6: Update μ'' by the gradient method in (5.40) and let $p_{m,n,c} = \iota_{m,n} p_c, \forall m, n$. If μ'' or ω_m is not converged, go to Step 5;

Step 7: If $\mathbf{p}_c$ is not converged, go to Step 4;

Output: $\mathbf{S}, \mathbf{P}$.

Up to now, we have already presented three algorithms to design non-orthogonal beams for massive access in the cellular IoT. As analyzed above, for the three algorithms, the degrees of freedom for the beam selection are decreased but the number of optimization variables is increased in sequence. Therefore, it is possible to choose a proper beam design algorithm according to the requirements of system performance and computational complexity.

5.4 Numerical Results

To validate the effectiveness of the proposed algorithms for non-orthogonal beamspace multiple access in the cellular IoT with massive connections, we carry out extensive numerical simulations in such a scenario: $N_t = 64$, $K = 50$. and $P_{\max} = 10\,\text{dB}$. The UEs are uniformly distributed in a circle with the BS as the centre and of the radius 50 m. Thus, the number of clusters is dynamically changed as the UEs move. We adopt a typical urban (TU) channel model for the cellular IoT according to the document of 3GPP TR 45.820 [24]. For ease of notation, we use SNR (in dB) to represent the term $10 \log_{10} P_{\max}$.

First, we compare the performance of the three proposed algorithms and the traditional single-beam algorithm in Fig. 5.6. Specifically, the traditional single-

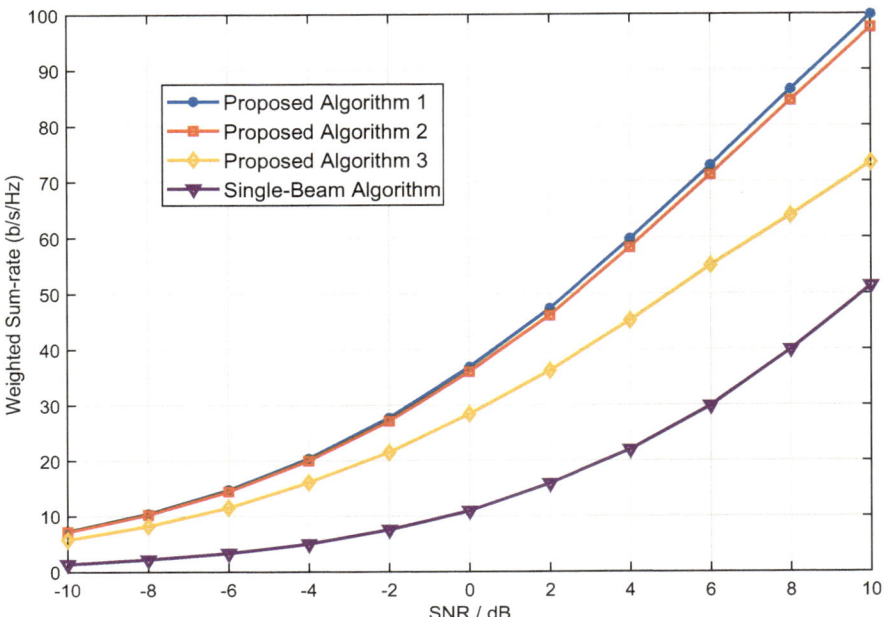

Fig. 5.6 Performance comparison of different beam design algorithms

beam algorithm uses the base beam specified by the AoD information as the transmit beam for a given cluster. It is seen that the three proposed multiple-beam combination algorithms perform much better than the single-beam algorithm, and the performance gain enlarges as the SNR increases. This is because the proposed algorithms have more degrees of freedom to design the transmit beams, then can mitigate the co-channel interference to a large extend. Therefore, the proposed algorithms have a strong capability to support massive access over limited radio spectrum. For the three proposed algorithms, Algorithm 2 can nearly achieve the same performance as Algorithm 1 in the whole SNR region. In other words, it is enough to construct the transmit beam based on a few base beams with high correlation. Since Algorithm 2 can obtain a balance between system performance and computational complexity, we take it as a typical non-orthogonal beamspace multiple access algorithm for performance comparison with the other multiple access technique in the following.

Then, we show the performance advantage of the proposed non-orthogonal beamspace multiple access scheme (Algorithm 2) over the traditional orthogonal multiple access scheme in Fig. 5.7. To be specific, the traditional orthogonal multiple access scheme is a time division multiple access (TDMA) scheme, which divides a time slot into K sub-slots and each UE occupies a sub-slot exclusively. The TDMA selects the optimal base beam as the transmit beam for maximizing the received SINR. It is seen that Algorithm 2 has an obvious performance gain over the TDMA.

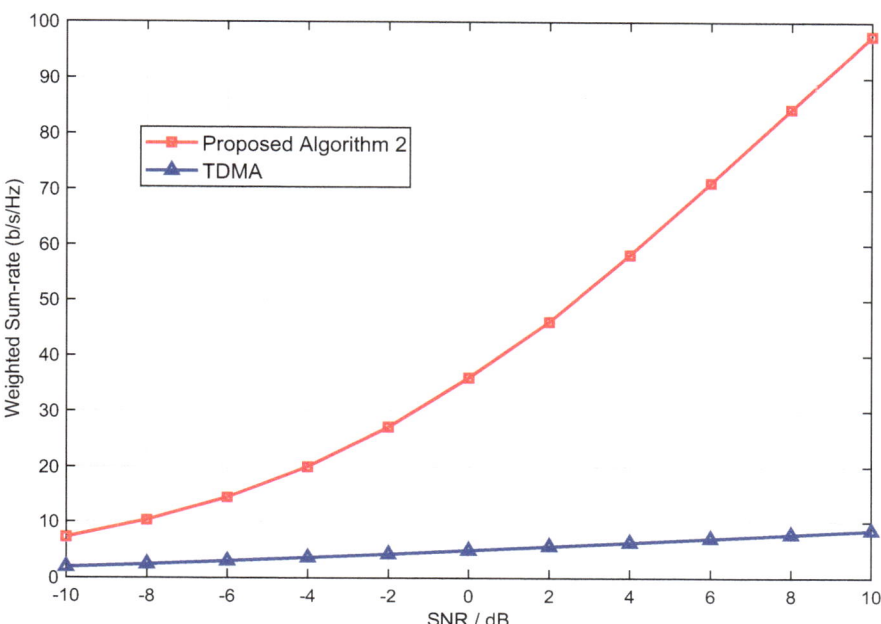

Fig. 5.7 Performance comparison of the NOMA and OMA schemes

This is because Algorithm 2 effectively exploits the spatial multiplexing capability offered by the multiple-antenna BS, and thus can significantly improve the sum rate. Although there is high co-channel interference in the high SNR region, Algorithm 2 is able to mitigate the interference by using beamspace beamforming. Hence, the performance gain becomes large as the SNR increases.

Figure 5.8 shows the influence of the number of UEs K on the sum rate when the number of BS antenna N_t is fixed. It is found that the performance of the three proposed algorithms improves as K increases. This is because more UEs can leverage the spatial multiplexing gain offered by the multiple-antenna BS. However, as K increases, the co-channel interference also sharply increases. Under this condition, Algorithm 1 and 2 have more degrees of freedom to mitigate the interference, and thus their performance improves continuously. For Algorithm 3, there exists high residual interference after beamspace beamforming at the BS and SIC at the UEs. As a result, the sum rate of Algorithm 3 increases slowly and becomes saturated fast.

Figure 5.9 investigates the impact of the number of BS antenna N_t on the sum rate for a given number of UEs. As is intuitively observed, a large number of BS antennas can provide a high angular resolution in the beamspace, and hence designs highly accurate transmit beams for interference mitigation and signal enhancement. As a result, the sum rates of the three proposed algorithms improve as the number of BS antennas increases. Therefore, we can improve the performance of massive

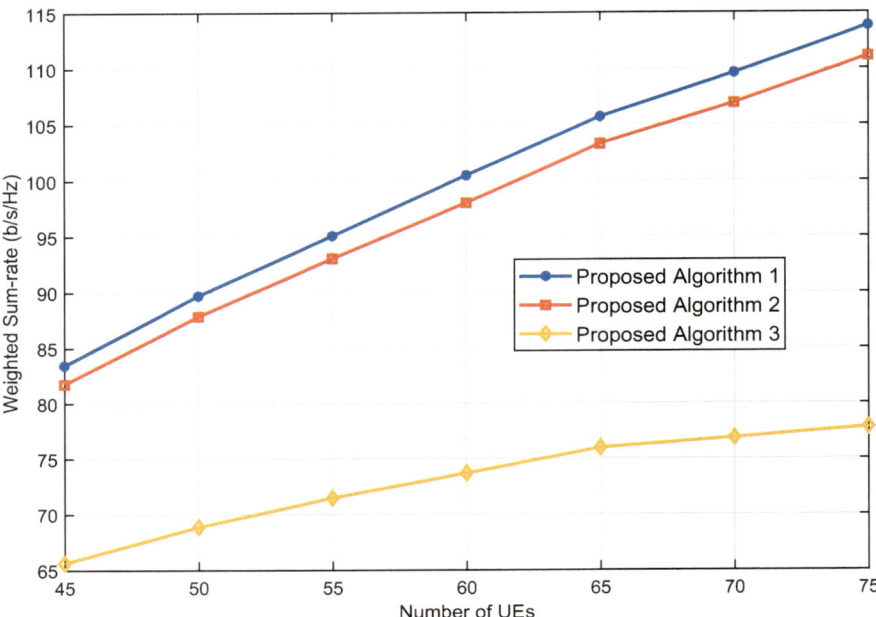

Fig. 5.8 The influence of the number of UEs on the sum rate

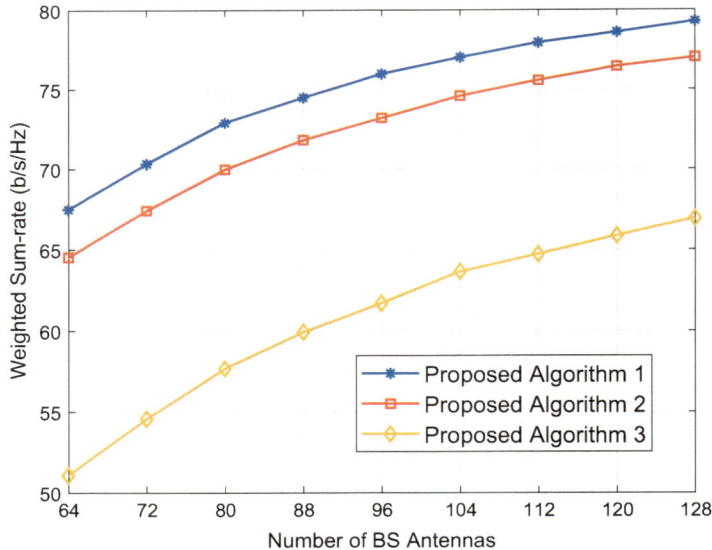

Fig. 5.9 The impact of the number of BS antennas on the sum rate

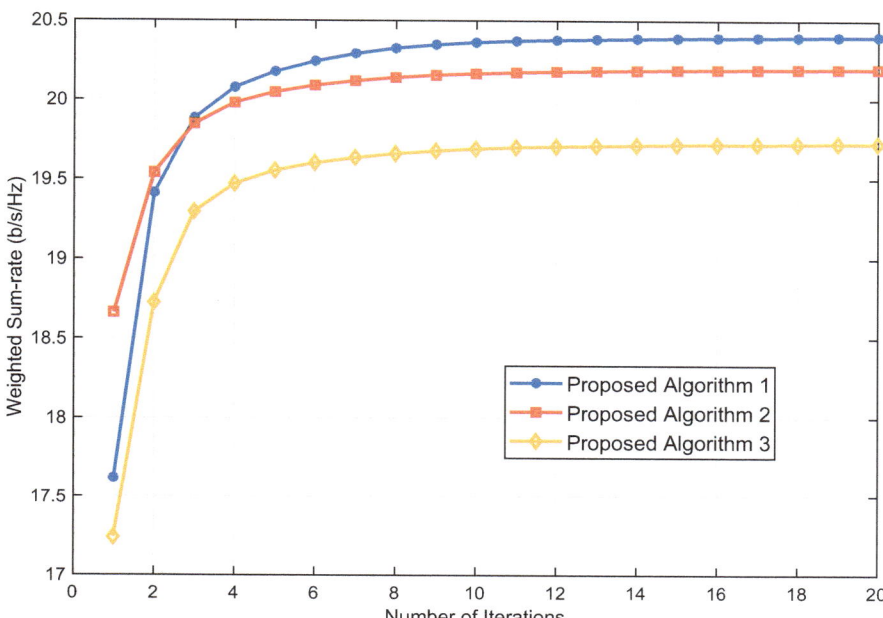

Fig. 5.10 The convergence behavior of the proposed three algorithms

access by simply adding the BS antennas, which is a major advantage of the 5G BS with a large-scale antenna array. However, it is found that the performance gain by adding the BS antennas becomes small gradually. This is because there is nearly one UE in a cluster when the angular resolution is large enough.

Finally, we check the convergence behavior of the three proposed algorithms in Fig. 5.10. It is seen that all the algorithms converge after no more than 10 times iterations. As discussed earlier, the beam design is a linear combination of multiple base beams with the power obtained by iterations. Thus, the proposed algorithms have low computational complexity and fast convergence behavior.

5.5 Conclusion

In this chapter, we designed a massive access framework for the cellular IoT by making use of the characteristics of beamspace. Especially, we proposed to adopt non-orthogonal transmit beams to improve the performance of massive access systems. Three simple but effective non-orthogonal beam design algorithms were presented from the perspectives of maximizing the performance upper bound.

Appendix   The Proof of Theorem 1

Prior to proving Theorem 1, we first provide the following lemma [11]:

Lemma 3 *If* **A**, **B**, *and* **X** *are symmetric positive semi-definite matrices, the matrix function* $f(\mathbf{X})$ *in below is concave with respect to* **X**.

$$f(\mathbf{X}) = \log_2 \det(\mathbf{I} + \mathbf{AX}) - \log_2 \det(\mathbf{I} + \mathbf{BX}). \tag{5.43}$$

According to the definition, the achievable rate of the $\text{UE}_{m,n}$ in (5.13) can be transformed as

$$
\begin{aligned}
r_{m,n} &= \log_2 \left(\frac{\sum\limits_{j=1,j\neq m}^{M} \sum\limits_{i=1}^{N_j} |\bar{\mathbf{h}}_{m,n}^H \mathbf{v}_{m,n,j,i}|^2 + \sum\limits_{i=1}^{n} |\bar{\mathbf{h}}_{m,n}^H \mathbf{v}_{m,n,m,i}|^2 + 1}{\sum\limits_{j=1,j\neq m}^{M} \sum\limits_{i=1}^{N_j} |\bar{\mathbf{h}}_{m,n}^H \mathbf{v}_{m,n,j,i}|^2 + \sum\limits_{i=1}^{n-1} |\bar{\mathbf{h}}_{m,n}^H \mathbf{v}_{m,n,m,i}|^2 + 1} \right) \\
&= \log_2 \det \left(\bar{\mathbf{h}}_{m,n} \bar{\mathbf{h}}_{m,n}^H \left(\sum\limits_{j=1,j\neq m}^{M} \sum\limits_{i=1}^{N_m} \mathbf{V}_{m,n,j,i} + \sum\limits_{i=1}^{n} \mathbf{V}_{m,n,m,i} \right) + \mathbf{I} \right) \\
&\quad - \log_2 \det \left(\bar{\mathbf{h}}_{m,n} \bar{\mathbf{h}}_{m,n}^H \left(\sum\limits_{j=1,j\neq m}^{M} \sum\limits_{i=1}^{N_m} \mathbf{V}_{m,n,j,i} + \sum\limits_{i=1}^{n-1} \mathbf{V}_{m,n,m,i} \right) + \mathbf{I} \right),
\end{aligned}
\tag{5.44}
$$

where $\mathbf{v}_{m,n,j,i} = \Lambda_{m,n}^{\frac{1}{2}} \mathbf{P}_{j,i}^{\frac{1}{2}} \mathbf{s}_{j,i}$ and $\mathbf{V}_{m,n,j,i} = \mathbf{v}_{m,n,j,i} \mathbf{v}_{m,n,j,i}^H$. Equation (5.44) holds true due to the fact that $\det(\mathbf{I}+\mathbf{AB}) = \det(\mathbf{I}+\mathbf{BA})$. Based on Lemma 3, $r_{m,n}$ is a concave function of $\|\bar{\mathbf{h}}_{m,n}\|^2$. Thus, applying the Jensen's inequality yields

$$
\begin{aligned}
E\{r_{m,n}\} &\leq \log_2 \det \left(E[\bar{\mathbf{h}}_{m,n} \bar{\mathbf{h}}_{m,n}^H] \left(\sum\limits_{j=1,j\neq m}^{M} \sum\limits_{i=1}^{N_m} \mathbf{V}_{m,n,j,i} + \sum\limits_{i=1}^{n} \mathbf{V}_{m,n,m,i} \right) + \mathbf{I} \right) \\
&\quad - \log_2 \det \left(E[\bar{\mathbf{h}}_{m,n} \bar{\mathbf{h}}_{m,n}^H] \left(\sum\limits_{j=1,j\neq m}^{M} \sum\limits_{i=1}^{N_m} \mathbf{V}_{m,n,j,i} + \sum\limits_{i=1}^{n-1} \mathbf{V}_{m,n,m,i} \right) + \mathbf{I} \right) \\
&= \log_2 \det \left(\sum\limits_{j=1,j\neq m}^{M} \sum\limits_{i=1}^{N_m} \mathbf{V}_{m,n,j,i} + \sum\limits_{i=1}^{n} \mathbf{V}_{m,n,m,i} + \mathbf{I} \right) \\
&\quad - \log_2 \det \left(\sum\limits_{j=1,j\neq m}^{M} \sum\limits_{i=1}^{N_m} \mathbf{V}_{m,n,j,i} + \sum\limits_{i=1}^{n-1} \mathbf{V}_{m,n,m,i} + \mathbf{I} \right).
\end{aligned}
\tag{5.45}
$$

Then, the upper bound of the weighted sum of the ergodic rates can be written as

$$
R_{ub} = \sum_{m=1}^{M}\sum_{n=1}^{N_m}\alpha_{m,n}\left(\log_2 \det\left(\sum_{j=1,j\neq m}^{M}\sum_{i=1}^{N_m}\mathbf{V}_{m,n,j,i} + \sum_{i=1}^{n}\mathbf{V}_{m,n,m,i} + \mathbf{I}\right)\right.
$$
$$
\left. - \log_2 \det\left(\sum_{j=1,j\neq m}^{M}\sum_{i=1}^{N_m}\mathbf{V}_{m,n,j,i} + \sum_{i=1}^{n-1}\mathbf{V}_{m,n,m,i} + \mathbf{I}\right)\right)
$$
$$
= \sum_{m=1}^{M}\sum_{n=1}^{N_m}\sum_{c=1}^{N_t}\alpha_{m,n}\left(\log_2\left(\left(\sum_{j=1,j\neq m}^{M}\sum_{i=1}^{N_j}s_{j,i,c}p_{j,i,c} + \sum_{i=1}^{n}s_{m,i,c}p_{m,i,c}\right)\eta_{m,n,c} + 1\right)\right.
$$
$$
\left. - \log_2\left(\left(\sum_{j=1,j\neq m}^{M}\sum_{i=1}^{N_j}s_{j,i,c}p_{j,i,c} + \sum_{i=1}^{n-1}s_{m,i,c}p_{m,i,c}\right)\eta_{m,n,c} + 1\right)\right), \qquad (5.46)
$$

where Eq. (5.46) follows the fact that $\mathbf{V}_{m,n,j,i}$ is a diagonal matrix. The proof completes.

References

1. Z. Wei, J. Yuan, D.W.K. Ng, E. Maged, Z. Ding, A survey of downlink non-orthogonal multiple access for 5G wireless communication networks. ZTE Commun. **14**(4), 17–25 (2016)
2. L. Dai, B. Wang, Y. Yuan, S. Han, I. Chih-Lin, Z. Wang, Non-orthogonal multiple access for 5G: solutions, challenges, opportunities, and future research trends. IEEE Commun. Mag. **53**(9), 74–81 (2015)
3. Z. Ding, Z. Yang, P. Fan, H.V. Poor, On the performance of non-orthogonal multiple access in 5G systems with randomly deployed users. IEEE Signal Process. Lett. **21**(12), 1501–1505 (2014)
4. J. Seo, Y. Sung, Beam design and user scheduling for nonorthogonal multiple access with multiple antennas based on pareto optimality. IEEE Trans. Signal Process. **66**(11), 2876–2891 (2018)
5. V.-D. Nguyen, H.D. Tuan, T.Q. Duong, H.V. Poor, O.-S. Shin, Precoder design for signal superposition in MIMO-NOMA multicell networks. IEEE J. Sel. Areas Commun. **35**(12), 2681–2695 (2017)
6. X. Chen, Z. Zhang, C. Zhong, D.W.K. Ng, Exploiting multiple-antenna for non-orthogonal multiple access. IEEE J. Sel. Areas Commun. **35**(10), 2207–2220 (2017)
7. H.V. Cheng, E. Björnson, E.G. Larsson, Performance analysis of NOMA in training-based multiuser MIMO systems. IEEE Trans. Wirel. Commun. **17**(1), 372–385 (2018)
8. X. Chen, Z. Zhang, C. Zhong, R. Jia, D.W.K. Ng, Fully non-orthogonal communication for massive access. IEEE Trans. Commun. **66**(4), 1717–1731 (2018)
9. X. Chen, Z. Zhang, H.-H. Chen, On distributed antenna system with limited feedback precoding-opportunities and challenges. IEEE Wirel. Commun. **17**(2), 80–88 (2010)
10. X. Chen, C. Yuen, Performance analysis and optimization for interference alignment over MIMO interference channels with limited feedback. IEEE Trans. Signal Process. **62**(7), 1785–1795 (2014)
11. C. Sun, X. Gao, S. Jin, M. Matthaiou, Z. Ding, C. Xiao, Beam division multiple access transmission for massive MIMO communications. IEEE Trans. Commun. **63**(6), 2170–2784

(2015)

12. L. You, X. Gao, G.Y. Li, X.-G. Xia, N. Ma, BDMA for millimeter-wave/terahertz massive MIMO transmission with per-beam synchronization. IEEE J. Sel. Areas Commun. **35**(7), 1550–1563 (2017)

13. J. Zhao, F. Gao, W. Jia, S. Zhang, S. Jin, H. Lin, Angular domain hybrid precoding and channel tracking for millimeter wave massive MIMO systems. IEEE Trans. Wirl. Commun. **16**(10), 6868–6880 (2017)

14. X. Chen, Z. Zhang, Exploiting angular domain information for precoder design in distributed antenna system. IEEE Trans. Signal Process. **58**(11), 5791–5801 (2010)

15. B. Wang, L. Dai, Z. Wang, N. Ge, S. Zhou, Spectrum and energy-efficient beamspace MIMO-NOMA for millimeter-wave communications using lens antenna array. IEEE J. Sel. Areas Commun. **35**(10), 2370–2382 (2017)

16. Y.I. Choi, J.W. Lee, M. Rim, C.G. Kang, On the performance of beam division nonorthogonal multiple access for FDD-based large-scale multi-user MIMO systems. IEEE Trans. Wirel. Commun. **16**(8), 5077–5089 (2017)

17. H. Lin, F. Gao, S. Jin, G.Y. Li, A new view of multi-user hybrid massive MIMO: non-orthogonal angular division multiple access. IEEE J. Sel. Areas Commun. **35**(10), 2268–2280 (2017)

18. J. Lin, W. Yu, N. Zhang, X. Yang, H. Zhang, W. Zhao, A survey on Internet of things: Architecture, enabling technologies, security and privacy, and application. IEEE Internet Things J. **4**(5), 1125–1142 (2017)

19. H. Bolcskei, M. Borgmann, A.J. Paulray, Impact of the propagation environment on the performance of space-frequency coded MIMO-OFDM. IEEE J. Sel. Areas Commun. **21**(3), 427–439 (2003)

20. A.J. Barabell, J. Capon, D.F. Delong, J.R. Johnson, K. Senne, Performance comparison of superresolution array processing algorithms. Technical report TST-72, Lincoln Laboratory, Massachusetts Institute of Technology, Cambridge (1984)

21. V. Raghavan, J.H. Kotecha, A.M. Sayeed, Why does the kronecker model result in misleading capacity estimates? IEEE Trans. Inf. Theory **56**(10), 4843–4864 (2010)

22. S.S. Christensen, R. Agarwal, E. de Carvalho, J.M. Cioffi, Weighted sum-rate maximization using weighted MMSE for MIMO-BC beamforming design. IEEE Trans. Wirel. Commun. **7**(12), 4792–4799 (2008)

23. Q. Shi, M. Razaviyayn, Z.-Q. Luo, C. He, An iteratively weighted MMSE approach to distributed sum-utility maximization for a MIMO interfering broadcast channel. IEEE Trans. Signal Process. **59**(9), 4331–4340 (2011)

24. 3GPP TR 45.820, Technical specification group GSM/EDGE radio access network; Cellular system support for ultra-low complexity and low throughput Internet of things (CIoT) (2015)

Chapter 6
Summary

Abstract This chapter presents a summary about massive access for the cellular IoT in 5G and beyond. In particular, we discuss the theories and techniques of massive access and their applications in the cellular IoT according to the characteristics of available CSI at the BS in different scenarios. Firstly, a massive access scheme for a fixed cellular IoT where full CSI is available at the BS is designed. Especially, spatial beam and transmit power are jointly optimized according to instantaneous CSI from the perspectives of maximizing the weighted sum rate and minimizing the total power consumption, respectively. Then, a low-mobility cellular IoT operated in FDD mode that partial CSI is obtained through a quantization codebook is studied, and the corresponding massive access scheme is provided by optimizing the feedback resource. Furthermore, a TDD mode-based cellular IoT is considered, and a fully non-orthogonal massive access scheme is proposed. To exploit the benefits of fully non-orthogonal massive access, the transmit power at both the BS and IoT devices is optimized. Finally, to satisfy the requirement of high mobility, a non-orthogonal beamspace massive access scheme is given, which can achieve a better spectral efficiency over fast-varying fading channels. Moreover, we analyze the challenging issues in the existing massive access schemes, and point out the future research directions for further improving the overall performance of the cellular IoT.

6.1 Concluding Remarks

With the increasing development of IoT, a massive number of IoT devices demand to access various wireless networks, so as to satisfy the requirements of a variety of typical applications, i.e., metropolitan time-frequency perception, virtual navigation/virtual management, smart traffic, and environmental monitoring. As the biggest wireless networks, the 5G and even the next-generation wireless network are required to provide wireless access with QoS guarantee for a great number of IoT applications. However, the currently used multiple access techniques are difficult

X. Chen, *Massive Access for Cellular Internet of Things Theory and Technique*,
SpringerBriefs in Electrical and Computer Engineering,
https://doi.org/10.1007/978-981-13-6597-3_6

to support massive access over limited radio spectrum. In this context, the non-orthogonal multiple access technique is widely regarded as a promising candidate for achieving spectral-efficient massive access. Especially, to resolve the two key problems in NOMA, namely severe co-channel interference and high computational complexity of SIC, the multiple-antenna techniques at the BS are usually combined with the NOMA in the scenario of massive access. As is well known that to exploit the benefits of multiple-antenna techniques, the BS should have accurate CSI. However, it is not a trivial task for the BS to obtain accurate CSI, especially in the scenario of massive access. In this book, we provide a few feasible solution for CSI acquisition according to the channel characteristics of the cellular IoT, and then design the corresponding massive access schemes. The main contributions of this book are summarized as follows.

In Chap. 1, we first introduce the characteristics of cellular IoT in 5G and beyond. In general, compared to traditional wireless networks, the cellular IoT has several new characteristics, e.g., a massive number of access devices, low latent and reliable wireless services, scarce wireless resources, and energy and capability-constrained wireless devices. These characteristics increase the difficulty in the design of wireless access, especially a few of characteristics are mutually contradictory. Then, based on these characteristics, we discuss several key techniques of massive access, including CSI acquisition, user clustering, superposition coding and SIC. Specifically speaking, the BS first obtains a certain kind of CSI about a massive number of IoT devices by some means, i.e., instantaneous and statistical CSI. Then, the BS partitions the IoT devices into multiple clusters for achieving a balance between system performance and computational complexity. Furthermore, the BS carries out superposition coding based on available CSI and user clustering, and broadcasts the coded signal over the downlink channels. Finally, the IoT devices perform SIC on the received signal to recover the desired information.

In Chap. 2, we consider a cellular IoT where the IoT devices are fixed. Thus, the channels vary very slowly, and the BS is able to obtain full CSI about the downlink channels of the IoT devices. Based on full CSI, the IoT devices are partitioned into multiple clusters according to their channel direction and gain. On the one hand, the nearly same channel direction within a cluster is beneficial to mitigate the inter-cluster interference and enhance the quality of the received signal. On the other hand, the distinct channel gain in a cluster facilitates to perform SIC. Then, based on available CSI and user clustering, the BS carries out superposition coding on the signals to be transmitted. The superposition coding includes two steps, one is power allocation within the cluster, and the other is spatial beamforming across the clusters. The power allocation within the cluster can coordinate the intra-cluster interference, while spatial beamforming may mitigate the inter-cluster interference. Finally, the BS broadcasts the coded signal over the downlink channels, and the IoT devices performs SIC within a cluster. Due to energy and size constraints, the IoT devices have a limited computational ability. Thus, the decoding error during the SIC is inevitable, resulting in residual interference after SIC. To alleviate the impact of imperfect SIC on the performance of massive access, we propose to jointly optimize the spatial beam and transmit power from the perspectives of maximizing the weighted sum rate and minimizing the total power consumption, respectively. Moreover, for further reducing the computational complexity, we provide two simplified massive access schemes adopting ZF beamforming at the BS fixedly. It has been shown that the proposed massive access schemes can effectively alleviate the impact of imperfect SIC, and thus enhance the robustness of the cellular IoT.

In Chap. 3, we focus on a cellular IoT where the IoT devices move with a low speed. In such a scenario, the channels are slowly time-varying. In other words, the channels remain constant within a time slot, and fade over time slots. We consider the cellular IoT operates in FDD mode, such that CSI should be conveyed from the IoT devices to the BS through quantized feedback. To be specific, each IoT device first obtains the CSI about the downlink channel through channel estimation, and quantizes the CSI by selecting an optimal codeword from a predetermined codebook. Then, the IoT device conveys the index of the optimal codewords with a finite number of bits to the BS, and thus the BS can recover the quantized CSI from the same codebook. Since the BS only has partial CSI, it carries out user clustering according to the position information of the IoT devices. In specific, the IoT devices in the nearly same physical direction but different access distances are grouped into a cluster. Furthermore, the BS directly conducts superposition coding based on imperfect CSI. Similarly, the signals within a cluster are weighted summed with the transmit power as the weighted coefficient, and then the signals of all clusters are weighted summed with the ZF beam as the weighted coefficient. Once the IoT devices receive the coded signal, SIC is adopted within a cluster to cancel partial intra-cluster interference. The number of quantization bits for CSI conveyance from the IoT device to the BS determines the CSI accuracy and thus the performance of the cellular IoT, but it is impossible to allocate a large number of feedback bits to each devices in the scenario of massive access due to a finite capacity of the feedback link. In this context, we propose to allocate the feedback bits according to channel conditions and system parameters, so as to obtain the satisfactory CSI for all devices with limited feedback resource. Moreover, we advocate to optimize the transmit power and transmission mode for further improving the overall performance of the cellular IoT. Simulation results validate the effectiveness of the proposed massive access based on quantized codebooks.

In Chap. 4, we also consider a cellular IoT with low-mobility IoT devices. Differently, the cellular IoT operates in TDD mode, and thus the BS can obtain the CSI by channel estimation directly. Generally speaking, the IoT devices send training sequences over the uplink channels, and the BS acquires the CSI about the uplink channels via channel estimation. Due to channel reciprocity in TDD mode, the uplink CSI can be used as the downlink CSI. In order to obtain accurate CSI, the training sequences are usually orthogonal of each other. In order to guarantee pairwise orthogonal, the length of training sequences should be larger than the number of IoT devices. However, in the scenario of massive access, the length of training sequences may become a large proportion of a time slot, resulting in a short duration for information transmission. In a worst case, the training sequence is longer than the channel coherent time. As a result, the estimated CSI is outdated. To solve this challenge, we propose a fully non-orthogonal communication framework for massive access. First, the channel estimation is non-orthogonal. The IoT devices in a cluster share the same training sequence, and the training sequences across the clusters are orthogonal. The sharing of training sequence sharply decreases the required number of training sequence, and then the length of training sequences is reduced. However, the non-orthogonal channel estimation may lead to the decreasing of CSI accuracy, and the CSI accuracy within a cluster are coupled. Second, the user access is also non-orthogonal. For the proposed fully non-orthogonal communication framework, there exists severe co-channel interference during the both stages of channel estimation and user access, resulting in a severe performance degradation. In order to achieve a spectral-efficient massive access for the cellular IoT, we propose to optimize the transmit power of the devices at the stage of channel estimation and the transmit power of the BS at the stage of user access accordingly. Numerical simulations show that the proposed non-orthogonal communication framework can support massive access with finite wireless resources.

In Chap. 5, we concentrate on a high-mobility cellular IoT, where the IoT devices move fast. Thus, the channels also vary so fast that it is impossible to feed back the CSI or send training sequences in each time slot. Meanwhile, it is necessary to design low-complexity superposition coding algorithm for reducing the processing delay over fast time-varying fading channels. In this context, we propose a non-orthogonal beamspace massive access scheme. In other words, we utilize beamspace CSI, namely statistical CSI, to design the massive access scheme. Since statistical CSI remains constant during a relatively long time, it is particularly applicable in the system in the fast time-varying environment. According to the characteristics of the beamspace, the base beams are asymptotically orthogonal. Thus, the whole beamspace are partitioned into multiple orthogonal subspaces, and each subspace relates to a base beam. For simplifying the user clustering, we group the IoT devices in a subspace into a cluster. Since the IoT devices in a cluster are randomly distributed in the subspace, the base beam cannot effectively enhance the performance of all the devices in a cluster. To this end, we propose to construct non-orthogonal transmit beam for the devices. First, we give a full-space multiple-beam design algorithm, which constructs an independent transmit beam for each device with all base beams. To reduce the design complexity, we then present a partial space design algorithm, which constructs an independent transmit beam for each device with partial base beams. For further decreasing the required number of transmit beams, we provide a partial-space single-beam design algorithm, which constructs a transmit beam for the devices in the same cluster with partial base beams. Since the transmit beam is a linear combination of the base beams, the superposition coding method has a low computational complexity. It has been shown that the proposed non-orthogonal beamspace massive access schemes perform better than the baseline schemes.

6.2 Future Works

Despite many fruitful research efforts in studying the theories and techniques of massive access in the cellular IoT, there are many challenging issues remained to be tackled. In the following, we list some initial ideas and research directions in future works.

First, this book aims to achieve a spectral-efficient massive access for the cellular IoT, so as to provide a data transmission with a large capacity for a massive number of wireless devices, even in the scenario of high mobility. In fact, other than large capacity, high mobility, and high peak data rate, the cellular IoT has several QoS requirements, i.e., ultra reliability, low latency, and low power consumption [1, 2]. However, these performance metrics might be coupled of each other. For instance, a large capacity may lead to a high power consumption, while the latency may increase in order to guarantee ultra reliability. In this context, on the one hard, it is necessary to apply new techniques to enhance the overall performance of the cellular IoT. As a simple example, the short-packet communication can be adopted to reduce the latency [3]. On the other hand, it is desired to provide differentiated QoS guarantee for various IoT applications. Especially, the human-type and machine-type communications have distinct requirements [4]. Therefore, these performance metrics should achieve a tradeoff for a specific IoT application. Thus, it is possible to fulfill the QoS requirements with limited wireless resources.

Second, the cellular IoT is a key component of 5G and beyond wireless networks. Thus, it makes sense to apply the 5G new radio (NR) techniques to further explore the potential of the cellular IoT. In this book, we have made an attempt to deploy the massive MIMO and NOMA techniques to enhance the performance of massive access. Actually, the two 5G key techniques have been proven by theoretical analysis, numerical simulation and practical measurement that they can support spectral-efficient massive access. In fact, there are many other promising 5G NR techniques which are suitable for the cellular IoT [5–7]. For instance, millimeter wave (mmWave) can provide a huge radio spectrum for short-distance communications [8]. Furthermore, by combining massive MIMO and mmWave techniques, it is likely to realize long-distance cellular communications for a massive number of IoT devices [9]. Moreover, the new 5G waveform techniques, e.g., filter bank multi-carrier (FBMC), universal filtered multi-carrier (UFMC), generalized frequency division multiplexing (GFDM), and filtered OFDM (F-OFDM) can significantly enhance the efficiency, reliability, and flexibility of the cellular IoT [10, 11].

Third, most of the IoT devices are energy-limited simple nodes. For some applications with low power consumption, the IoT devices can work for a long time. However, in some applications that require frequently signal processing and data transmission, the energy is usually in shortage. To support advanced IoT applications, the battery of the IoT device should be charged or replaced frequently. However, the frequent battery charging and replacement leads to a prohibitive cost when there are a massive number of IoT devices. Moreover, in some extreme applications, e.g., underwater measurement, the battery is difficult to be replaced.

In summary, the traditional wireline charging method is inefficient for the cellular IoT with a massive number of devices. Recently, wireless power transfer is widely regarded as an enabling charging technique for the cellular IoT [12–14]. The IoT devices harvest the electromagnetic energy from the radio frequency (RF) signals directly, and then transform it as the electric energy [15]. Since the RF signal is controllable, wireless power transfer can provide a reliable energy supply. Especially, the RF signal can carry both information and energy, thus it is possible to realize completely wireless communication, including wireless charging and wireless transmission [16]. However, wireless power transfer may suffer from path loss and channel fading, resulting in a low efficiency. In order to wirelessly charge a massive number of devices, it is imperative to design highly-efficient methods of wireless power transfer according to the characteristics of the cellular IoT [17].

Forth, the cellular IoT can provide a variety of wireless applications, and the data of partial applications is confidential [18]. For instance, the personal information should be submitted in some applications. However, due to the broadcast nature of wireless channels, the wireless signal can be received by any node, resulting in a not-small probability of information leakage. In conventional wireless communications, information security is guaranteed by using the upper-layer encryption techniques [19]. With the fast development of IT techniques, the encryption techniques become very complicated. As mentioned above, the IoT devices are energy and capability-constrained nodes, which are not able to run complicated encryption algorithms. Recently, physical layer security, as a complimentary of the upper-layer encryption, has received considerable interests. In general, physical layer security makes use of the random characteristics of physical channels, i.e., fading, noise, and interference, to make the capacity of the legitimate channel greater than that of the eavesdropper channel [20, 21]. Hence, the eavesdropper is not capable of decoding the received signal correctly. Due to the appealing characteristics of physical layer security, it is more suitable to the cellular IoT for providing secure communications. However, in the cellular IoT with massive connections, the received signal may suffer from severe co-channel interference, which degrades the performance of physical layer security [22, 23]. Therefore, it is necessary to design the physical layer security based on the features of the cellular IoT.

Finally, this book provides several feasible and effective solutions for the cellular IoT in the downlink communication. In other words, we only solve one side of the problem of massive access. The other side of the problem of massive access, namely the uplink communication, has not been well studied. It is intuitive that the uplink case has its unique challenging issues. For example, the user activity should be detected before information transmission in the uplink communications [24, 25]. In other words, the BS needs to know the state of the IoT devices. Since activity detection is a precondition of uplink massive access, it is a must to guarantee a reliable activity detection. Moreover, there are many open issues in the uplink massive access.

References

1. M.R. Palattella, M. Dohler, A. Grieco, G. Rizzo, J. Torsner, T. Engel, L. Ladid, Internet of things in the 5G era: enablers, architecture, and business models. IEEE J. Sel. Areas Commun. **34**(3), 510–527 (2016)
2. M. Agiwal, A. Roy, N. Saxena, Next generation 5G wireless networks: a comprehensive survey. IEEE Commun. Surys. Tuts **18**(3), 1617–1655 (2016)
3. G. Durisi, T. Koch, P. Popovski, Toward massive, ultrareliable, and low-latency wireless communication with short packet. Proc. IEEE **104**(9), 1711–1726 (2016)
4. C. Bockelmann, N. Pratas, H. Nikopour, K. Au, T. Svensson, C. Stefanovic, P. Popovski, A. Dekorsy, Massive machine-type communications in 5G: physical and MAC-lay solutions. IEEE Commun. Mag. **54**(9), 59–65 (2016)
5. J.G. Andrews, S. Buzzi, W. Choi, S.V. Hanly, A. Lozano, A.C.K. Soong, J.C. Zhang, What will 5G be? IEEE J. Sel. Areas Commun. **32**(6), 1065–1082 (2014)
6. C.-X. Wang, F. Haider, X. Gao, X.-H. You, Y. Yang, D. Yuan, H.M. Aggoune, H. Haas, S. Fletcher, E. Hepsaydir, Cellular architecture and key technologies for 5G wireless communication networks. IEEE Commun. Mag. **52**(2), 122–130 (2014)
7. V.W.S. Wong, R. Schober, D.W.K. Ng, L.-C. Wang, *Key Technologies for 5G Wireless Systems* (Cambridge University Press, Cambridge, 2017)
8. T.S. Rappaport, S. Sun, R. Mayzus, H. Zhao, Y. Azar, K. Wang, G.N. Wong, J.K. Schulz, M. Samimi, F. Gutierrez, Millimeter wave mobile communications for 5G cellular: it will work! IEEE Access **1**, 335–349 (2013)
9. W. Roh, J.-Y. Seol, J. Park, B. Lee, J. Lee, Y. Kim, J. Cho, K. Cheun, F. Aryanfar, Millimeter-wave beamforming as an enabling technology for 5G cellular communications: theoretical feasibility and prototype results. IEEE Commun. Mag. **52**(2), 106–113 (2014)
10. S.-Y. Lien, S.-L. Shieh, Y. Huang, B. Su, Y.-L. Hsu, H.-Y. Wei, 5G new radio: waveform, frame structure, multiple access, and initial access. IEEE Commun. Mag. **55**(6), 64–71 (2017)
11. B. Farhang-Boroujeny, H. Moradi, OFDM inspired waveforms for 5G. IEEE Commun. Surys. Tuts. **18**(4), 2474–2492 (2016)
12. P. Kamalinejad, C. Mahapatra, Z. Sheng, S. Mirabbasi, V.C.M. Leung, Y.L. Guan, Wireless energy harvesting for the Internet of things. IEEE Commun. Mag. **53**(6), 102–108 (2015)
13. X. Chen, Z. Zhang, H.-H. Chen, H. Zhang, Enhancing wireless information and power transfer by exploiting multi-antenna techniques. IEEE Commun. Mag. **53**(4), 133–141 (2015)
14. D. Mishra, G.C. Alexandropoulos, S. De, Energy sustainable IoT with individual QoS constraints through MISO SWIPT multicasting. IEEE Internet Things J. **5**(4), 2856–2867 (2018)
15. C. Zhong, X. Chen, Z. Zhang, G. Karagiannidis, Wireless powered communications: performance analysis and optimization. IEEE Trans. Commun. **63**(12), 5178–5190 (2015)
16. X. Chen, C. Yuen, Z. Zhang, Wireless energy and information transfer tradeoff for limited feedback multi-antenna systems with energy beamforming. IEEE Trans. Veh. Technol. **63**(1), 407–412 (2014)
17. X. Chen, X. Wang, X. Chen, Energy-efficient optimization for wireless information and power transfer in large-scale MIMO systems employing energy beamforming. IEEE Wirel. Commun. Lett. **2**(6), 667–670 (2013)
18. J. Granjal, E. Monteiro, J.S. Silva, Security for the internet of things: a survey of existing protocols and open research issues. IEEE Commun. Surys. Tuts. **17**(3), 1294–1312 (2015)
19. S.L. Keoh, S.S. Kumar, H. Tschofenig, Securing the Internet of things: a standardization perspective. IEEE Internet Things J. **1**(3), 265–275 (2014)
20. X. Chen, H.-H. Chen, Physical layer security in multi-cell MISO downlink with incomplete CSI-A unified secrecy performance analysis. IEEE Trans. Signal Process. **62**(23), 6286–6297 (2014)

21. Y. Wu, A. Khisti, C. Xiao, G. Caire, K.-K. Wong, X. Gao, A survey of physical layer security techniques for 5G wireless networks and challenges ahead. IEEE J. Sel. Areas Commun. **36**(4), 679–695 (2018)

22. X. Chen, L. Lei, H. Zhang, C. Yuen, Large-scale MIMO relaying techniques for physical layer security: AF or DF? IEEE Trans. Wirel. Commun. **14**(9), 5135–5146 (2015)

23. X. Chen, D.W.K. Ng, W. Gerstacker, H.-H. Chen, A survey on multiple-antenna techniques for physical layer security. IEEE Commun. Survs. Tuts. **19**(2), 1027–1053 (2017)

24. L. Liu, W. Yu, Massive connectivity with massive MIMO-part I: device activity detection and channel estimation. IEEE Trans. Signal Process. **66**(11), 2933–2946 (2018)

25. Z. Chen, F. Sohrabi, W. Yu, Sparse activity detection for massive connectivity. IEEE Trans. Signal Process. **66**(7), 1890–1904 (2018)